T0258169

Encyclopedia of Quantum Mechanics: Developments

Volume VII

Encyclopedia of Quantum Mechanics: Developments

Volume VII

Edited by **Ian Plummer**

New York

Published by NY Research Press,
23 West, 55th Street, Suite 816,
New York, NY 10019, USA
www.nyresearchpress.com

Encyclopedia of Quantum Mechanics: Developments
Volume VII
Edited by Ian Plummer

International Standard Book Number: 978-1-63238-162-0 (Hardback)

Contents

Preface

The advancement of quantum mechanics has given physics a completely new direction from that of classical physics in the early days. In fact, there is a constant development in this subject of a very fundamental nature, such as implications for the foundations of physics, physics of entanglement, geometric phases, gravity and cosmology and elementary particles as well. This book will be an important resource for researchers with respect to present topics of research in this developing area. The book has two broad sections: Quantization and Entanglement, and Quantum Information and Related Topics.

Various studies have approached the subject by analyzing it with a single perspective, but the present book provides diverse methodologies and techniques to address this field. This book contains theories and applications needed for understanding the subject from different perspectives. The aim is to keep the readers informed about the progresses in the field; therefore, the contributions were carefully examined to compile novel researches by specialists from across the globe.

Indeed, the job of the editor is the most crucial and challenging in compiling all chapters into a single book. In the end, I would extend my sincere thanks to the chapter authors for their profound work. I am also thankful for the support provided by my family and colleagues during the compilation of this book.

Editor

Quantization and Entanglement

Quantum Dating Market

C. M. Arizmendi and O. G. Zabaleta

Additional information is available at the end of the chapter

1. Introduction

Quantum algorithms have proven to be faster than the fastest known classical algorithms. Clearly, such a superiority means counting on a real quantum computer. Although this essential constraint elimination is in development process, many people is working on that and interesting advances are being made [1-3]. Meanwhile, new algorithms and applications of the existing ones are current research topics [4, 5]. One of the main goals of quantum computing is the application of quantum techniques to classical troubleshooting: the Shor algorithm [6], for example, is a purely quantum-mechanical algorithm which comes to solve the classical factoring problem, also the contribution of Lov Grover [7, 8] to speed up the search for items in an N-item database is very important. Both mathematical finds are the cornerstones of quantum computation, so, considerable amount of work on diverse subjects make use of them. Other algorithms which has been very important for quantum computing progress are Simon's and Deutsch-Jozsa's. Through the quantum games, Meyer in [9] and Eisert in [10], among other, showed that quantum techniques are generalizations of classical probability theory, allowing effects which are impossible in a classical setting. These and many other examples, show that there is no contradiction in using quantum techniques to describe non-quantum mechanical problems and solve hard to solve problems with classical tools. Adding, decision theory and game theory, two examples where probabilities theory is applied, deal with decisions made under uncertain conditions by real humans. Basically, the former considers only one agent and her decisions meanwhile the other considers also the conflicts that two or more players cause to each other through the decisions they take. Due to their inherent complexity this kind of problems results convenient to be analyzed by mean of quantum games models.

Widely observed phenomena of non-commutativity in patterns of behavior exhibited in experiments on human decisions and choices cannot be obtained with classical decision theory [11] but can be adequately described by putting quantum mechanics and decision theory together. Quantum mechanics and decision theory have been recently combined

[11–13] to take into account the indeterminacy of preferences that are determined only when the action takes place. An agent is described by a state that is a superposition of potential preferences to be projected onto one of the possible behaviors at the time of the interaction. In addition to the main goal of modeling uncertainty of preferences that is not due to lack of information, this formalism seems to be adequate to describe widely observed phenomena of non-commutativity in patterns of behavior.

Within this framework, we study the dating market decision problem that takes into account progressive mutual learning [14, 15]. This problem is a variation on the Stable Marriage Problem introduced by Gale and Shapley almost four decades ago [16], that has been recently reformulated in a partial information approach [17, 18]. Specifically, perfect information supposition is very far from being a good approximation for the dating market, in which men and women repeatedly go out on dates and learn about each other.

The dating market problem may be included in a more general category of matching problems where the elements of two sets have to be matched by pairs. Matching problems have broad implications in economic and social contexts [19, 20]. As possible applications one could think of job seekers and employers, lodgers and landlords, men and women who want to date, or solitary ciliates *courtship rituals* [21]. In our model players earn an uncertain payoff from being matched with a particular person on the other side of the market in each time period. Players have a list of preferred partners on the other set. Quantum exploration of partners is compared with classical exploration at the dating set. Nevertheless dating is not just finding, but also being accepted by the partner. The preferences of the chosen partner are important in quantum and classic performances.

Recently [22], we introduced a quantum formulation for decision matching problems, specifically for the dating game that takes into account mutual progressive learning. This learning is accomplished by representing women with quantum states whose associated amplitudes must be modified by men's selection strategies, in order to increase a particular state amplitude and to decrease the others, with the final purpose to achieve the best possible choice when the game finishes. Grover quantum search algorithm is used as a playing strategy. Within the same quantum formulation already used in [22], we will concentrate first on the information associated to the dating market problem. Since we deal with mixed strategies, the density matrix formalism is used to describe the system. There exists a strong relationship between game theories, statistical mechanics and information theories. The bonds between these theories are the density operator and entropy. From the density operator we can construct and understand the statistical behavior about our system by using statistical mechanics. The dating problem is analyzed through information theory under a criterion of maximum or minimum entropy. Even though the decisions players make are based on their payoffs, past experiences, believes, etc., we are not interested in that causes but in the consequences of the decision they take, that is, the influence of the strategies they apply on the quantum system stability. In order to identify the conditions of stability we will use the equivalence between maximum entropy states and those states that obey the Collective Welfare Principle that says that a system is stable only if it maximizes the welfare of the collective above the welfare of the individual [23].

Interesting properties merge when entanglement is considered in quantum models of social decision problems [24]. People decisions are usually influenced by other people actions, opinions, or beliefs, to the extent that they may proceed in ways that they would rarely or never do if moved by their own benefit. Love, hate, envy, or a close friendship, which encase a bit of everything, are examples of relationships between people that may correlate their decisions. So, as driven by a no local force, people may make an inconvenient choice in the heat of a competence. In order to formulate in a mathematical way this sort of problem we remodel the quantum dating between men and women with the inclusion of quantum entanglement between men decision states.

The chapter organization is as follows: First of all, to ease game theory unfamiliar readers comprehension a brief introduction to game basics is presented. In the course of the next sections the quantum dating game is particularly studied. In section 3 the Grover quantum search algorithm as a playing strategy is analyzed. In section 5, the system stability is under study. Finally, section 6 explores entangled strategies performance. At the end of each section the results and the consequent section discussions are set. The chapter ends with a final conclusion.

2. Quantum games

Game theory [25] is a collection of models (games) designed to study competing agents (players) decisions in some conflict situation. It tries to understand the birth and development of conflicting or cooperative behaviors among a group of individuals who behave rationally and strategically according to their personal interests. Although the theory was conceived in order to analyze and solve social and economy problems, existing applications go beyond [26]. Furthermore, the models reach not only individuals but also governments conflicts, institutions trades or smart machines (phones, computers) access management.

Before starting to explain quantum games basics, the classic games notation is presented. The game can be set in strategic (or normal) form or in extensive form, in any of them it has three elements: a set of players $i \in \mathcal{J}$ which is taken to be a finite set $0, ..., N - 1$, the set of pure strategies $S_i = \{s_0, s_1, ..., s_{N-1}\}$, $i = 0, ..., N - 1$ which is the set of all strategies available to the player, and the payoffs function $u_i(s_0, s_1, ..., s_{N-1})$, $i = 0, ..., N - 1$, where $s_i \in S_i$. In the strategic form, the game can be denoted by $G(N, S, u)$, where $S = S_0 \times S_1 \times ... \times S_{N-1}$ and $u = u_0 \times u_1 \times ... \times u_{N-1}$. Extensive form representation is useful when it is wanted to include not only who makes the move but also when the move is made. Players apply pure strategies when they are certain of what they want, but such condition is not always possible, so mixed strategies must be considered. A mixed strategy is a probability distribution over S which corresponds to how frequently each move is chosen.

As an example, we can mention the well-known *Prisoners Dilemma (PD)* : Two suspected of committing a crime are caught by the police. As there is insufficient evidence to condemn them, the police place the suspects into separate rooms to convince them to confess. If one of the prisoners confesses, and help the police to condene his partner, he gains his freedom and the other prisoner must serve of 10 years. But if both confess, they must serve a sentence of 3 years. In other case, if both refuse to confess, they both will be convicted of a lesser charge and will have to serve a sentence of only one year in prison. In summary, they can choose between two possible strategies "Confess" (C) or "Not Confess" (N). However, observe that the luck of each player depends both on his election as that of the other. As consequence, confessing is a dominant strategy because regardless the other player decision the one who chooses it avoid the worst conviction. The prisoners know that if neither confesses they must serve a minimum sentence. However, as no one knows the other strategy to do not confess is very risky, specially because camaraderie is not a common quality between criminals. It is very common to represent in a bimatrix the possible strategies combinations with their respective reward. The corresponding bimatrix for the prisoners game is 1.

$S_1 \setminus S_2$	C	N
C	3,3	0,10
N	10,0	1,1

Table 1. Prisoners Dilemma: C ≡ confess; N ≡ do not confess. The number on the left is for the years the prisoner S_1 prisoner must serve.

Quantum game theory is a classic game theory generalization. That is, quantum game strategies and outcomes include the classical as particularities, but also quantum features let the application of new strategies which leads to solutions classically imposible. The N players quantum game si denoted by $G(N, \mathcal{H}, \rho, S(\mathcal{H}), u)$, where $\mathcal{H}$ is the Hilbert space of the physical system and $\rho \in S(\mathcal{H})$ is the system initial condition, being $S(\mathcal{H})$ the associated space state. In quantum games, players strategies are represented by unitary operators, which in quantum mechanics are also known as evolution operators related to the system's Hamiltonian [27]. If we call U_i the operator corresponding to player i strategy, the N-players strategies operator results $U = U_0 \otimes U_1 ... U_i \otimes ... \otimes U_{N-1}$. Starting from the initial pure state $|\Psi_0\rangle$ of the system, players apply their strategies U in order to modify it according to their preferences, that is modifying the probability amplitudes associated with each base state. As a consequence, evolution from the initial system state to some state $|\Psi_1\rangle$ is given by $|\Psi_1\rangle = U|\Psi_0\rangle$. Quantum games provide new ways to cooperate, to eliminate dilemmas, and as a consequence new equilibriums arise. As can be seen in [10], for example, the dilemma is avoided in the quantum Prisoner's game. That is, the system equilibrium is not longer (C,C) to be (N,N).

3. Quantum search strategy

In the classic dating market game [28, 29], men choose women simultaneously from N options, looking for those women who would have some "property" they want. Unlike the traditional game, in the quantum version of the dating game, players get the chance to use quantum techniques, for example they can explore their possibilities using a quantum search algorithm. Grover algorithm capitalizes quantum states superposition characteristic to find some "marked" state from a group of possible solutions in considerably less time than a classical algorithm can do [8]. That state space must be capable of being translatable, say to a graph G where to find some particular state which has a searched feature or distinctive mark, throughout the execution of the algorithm. By "distinctive mark" we mean problems whose algorithmic solution are inspired by physical processes. Furthermore it is possible to guarantee that the searched node is marked by a minimum (maximum) value of a physical property included in the algorithm.

Let agents be coded as Hilbert space base states. As a result, men are able to choose from N_w women set $W = \{|0\rangle, |1\rangle, ..., |N_w - 1\rangle\}$. Table 1 displays four women states in the first column and some feature that makes them unique in the second column which we will code with a letter for simplicity.

If a player is looking for a woman with a feature "d", the table must be searched on its second column and when the desired "d" is found, look at the first column where the corresponding

woman	feature
$\lvert 0 \rangle$	a
$\lvert 1 \rangle$	b
$\lvert 2 \rangle$	c
$\lvert 3 \rangle$	d

Table 2. Sample woman database. Left column contains women states and right column displays a letter representing some feature or a feature set that characterizes each woman on the left.

chosen woman state is: $\lvert 3 \rangle$ in this example. The procedure is very simple if the table has just a few rows, but when the database gets bigger, the table in the best case would have to be entered $N_w/2$ times [30]. Under this framework we propose to use Grover algorithm in order to achieve man's decision in less time. Without losing generality let $N_w = 2^n$ being n the qubits needed to code N_w women. Quantum states transformation are made by applying Hilbert space operators U to them, following $\Psi_1 = U_1\Psi_0$ is a new system state starting from Ψ_0. As a consequence any quantum algorithm can be thought as a set of suitable linear transformations. Grover algorithm starts with n qubits in $\lvert 0 \rangle$, resulting $\psi_{ini} = \lvert 00..00 \rangle \equiv \lvert 0 \rangle^{\otimes n}$ the system initial state, where $\otimes$ symbol denotes Kronecker tensor product. Initially, the woman identified by state $\lvert 0 \rangle$ is chosen with probability one. The next step is to create superposition states and like many other quantum algorithms Grover uses Hadamard transform to do this task since it maps n qubits initialized with $\lvert 0 \rangle$ to a superposition of all n orthogonal states in the $\lvert 0 \rangle$, $\lvert 1 \rangle$,.. $\lvert n-1 \rangle$ basis with equal weight, $\psi_1 = H\psi_{ini} = \frac{1}{\sqrt{N_w}} \sum_{i=0}^{N_w - 1} \lvert i \rangle$. As an example, when $N_w = 4$, the state results $\psi_1 = H\lvert 00 \rangle = \frac{1}{2} \sum_{i=0}^{3} \lvert i \rangle = \frac{\lvert 00 \rangle + \lvert 01 \rangle + \lvert 10 \rangle + \lvert 11 \rangle}{2}$. One-qubit Hadamard transform matrix representation is (1), and n-qubits extension is $H^{\otimes n}$, see [27],

$$H = \frac{1}{\sqrt{2}} \begin{pmatrix} 1 & 1 \\ 1 & -1 \end{pmatrix} \tag{1}$$

Another quantum search algorithms characteristic, is the "Oracle", which is basically a black box capable of marking the problem solution. We call U_f the operator which implement the oracle

$$U_f(\lvert w \rangle \lvert q \rangle) = \lvert w \rangle \lvert q \oplus f(w) \rangle, \tag{2}$$

where $f(w)$ is the oracle function which takes the value 1 if w correspond to the searched woman, $f(w) = 1$, and if it is not the case it takes the value 0, $f(w) = 0$. The value of $f(w)$ on a superposition of every possible input w may be obtained [27]. The algorithm sets the target qubit $\lvert q \rangle$ to $\frac{1}{\sqrt{2}}(\lvert 0 \rangle - \lvert 1 \rangle)$. As a result, the corresponding mathematical expression is:

$$\lvert w \rangle \left(\frac{\lvert 0 \rangle - \lvert 1 \rangle}{\sqrt{2}} \right) \overset{U_f}{\longmapsto} (-1)^{f(w)} \lvert w \rangle \left(\frac{\lvert 0 \rangle - \lvert 1 \rangle}{\sqrt{2}} \right) \tag{3}$$

Observe that the second register is in an eigenstate, so we can ignore it, considering only the effect on the first register.

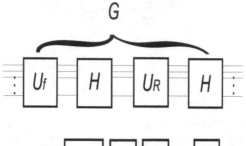

Figure 1. Grover Iteration

Figure 2. Grover Quantum searching algorithm

$$|w\rangle \xmapsto{U_f} (-1)^{f(w)}|w\rangle \tag{4}$$

Consequently, if $f(w) = 1$ a phase shift is produced, otherwise nothing happens. As we already stated our algorithm is based on the classical Gale-Shapley (GS) algorithm which assigns the role of proposers to the elements of one set, the men say, and of judges to the elements of the other.

Actually, for a more symmetric formulation of the algorithm where both sets are, at the same time, proposers and judges, it would be necessary another oracle which evaluates women features matching by means of another function $g(x)$ [31], but we will not go into that. As far as we are concerned up to now the Oracle is a device capable of recognizing and "mark" a woman who has some special feature, said hair color, money, good manners, etc. Oracle operator U_f makes one of two central operations comprising of a whole operation named Grover iterate G (Fig.1), and a rotation operator U_R, or conditional phase shift operator represented by equation (5).

U_R and U_f, together with Hadamard transformations represented by H blocks (1), in the order depicted by (Fig. 1), make the initial state vector asymptotically going to reach the solution state vector amplitudes. The symbol I in U_R equation is the identity operator.

$$U_R = 2|0\rangle\langle 0| - I \tag{5}$$

Furthermore, after applying Grover iterate, G, $O(\sqrt{N_w})$ times, the man finds the woman he is looking for. In Figure 1 Grover iterate is shown and Grover quantum algorithm scheme is depicted in 2.

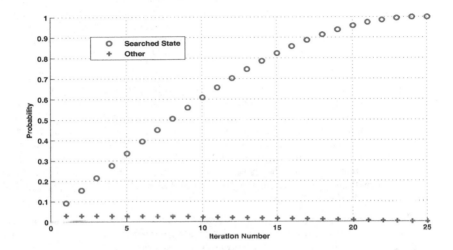

Figure 3. Evolution of the probability to find the chosen woman and the probability to find other woman as a function of the iteration number with Grover's algorithm.

As the number of iterations the algorithm makes depends on the size of the options set, this must be known at the beginning of simulations. Every operator has its matrix representation to be used in simulations. We suppose the player chooses a woman who has some specific particularity that would distinguish her from any other of the group, so we construct matrix U_f and other matrixes for that purpose. The evolution of the squared amplitude with the iteration number is shown in Figure 3. The searched state amplitude is initially the same for all possible states $|i>$ in the Ψ_1 expression. The fast increasing of the probability to find the preferred state on each iteration contrasts with the decreasing of the probability to find every other state. The example displayed is for $N_w = 1024$ women and as the can be seen in Figure 3, the number of iterations needed to get certainty to find the preferred woman are 25. Classically, a statistical algorithm would need approximately $N_w = 1024$ iterations.

Thus when a given man who wants to date a N_w size set selected woman, he must set his own U_f operator out, according to his preferences, and then let the algorithm do the job. The case of N_m men may be obtained generalizing the single man case: every one of them must follow the same steps. Nevertheless, achieving top choice is hard because of competition from other players and your dream partner may not share your feelings. If all players play quantum, the time to find woman is not an issue and the N stable solutions will be the same as for the classic formulation [32].

4. Quantum vs classic

To compare the quantum approach efficiency with the classical one we will consider some players playing quantum and others playing classic. Let us follow the evolution of agents representative from each group, Q and C respectively.

Q, that plays quantum can keep his state as a linear combination of all the prospective results when unitary transforms such as the described above for Grover's algorithm are applied, provided no measurement producing collapse to any of them is done. On the other hand, the only way C has to search such a database is to test the elements sequentially against the condition until the target is found. For a database of size N_w, this brute force search requires an average of $O(N_w/2)$ comparisons [7].

Two different games where both men want to date with the same woman are presented: In the first one player Q gives player C the chance to play first and both have only one attempt per turn, which means only one question to the oracle. The second game, in order that Q plays handicapped, is set out in the way that C can play $N_w/2$ times while Q only once, and player C plays first again. For the last case we analyzed two alternatives for the classic player: in the first one he plays without memory of his previous result and therefore, in every try he has $1/N_w$ probability to find the chosen woman to date, the other alternative permits the classic player to discard previous unfavorable outcomes at any try in order to avoid choosing them again and diminish the selection universe.

The player who invites the chosen woman first has more chances to succeed, as well as that who asks the same woman more times. Nevertheless the woman has the last word, and therefore the dating success for each player depends on that woman preferences. So, let us define P_c^i as the probability that woman i accepts dating the classic player C and P_q^i as the probability that she accepts the quantum player Q proposal. In order to compare performances, we consider $T = 1000$ playing times on turns and count the dating success times, then calculate the mean relative difference between Q and C success total number as $D/T = \frac{Qsuccess - Csuccess}{T}$, for different woman acceptation probabilities.

Initially, both players begin with the system in the initial state $\psi_1 = \frac{1}{\sqrt{N_w}} \sum_{i=0}^{N_w-1} |i\rangle$, therefore the probability to select any woman is the same for both, $p(w_i) = 1/N$. In the next step the Oracle marks one of the prospective women state according men preferences.

The results are highly dependent on the women set size N_w because, as mentioned above, Grover algorithm needs $O(\sqrt{(N_w)})$ steps to find the quantum player's chosen partner while the classic player must use $O(N_w)$ for the same task. In the case of only one woman and one man, for example, classic and quantum will not have any advantage on searching and the dating success difference for the first game will depend only on that woman preferences, that is, if $P_c > P_q$ then $D/T < 0$ and the quantum player will do better when $P_q > P_c$. Similar chances for both players is not usual in most quantum games, such as, for example the coin flip game introduced by Meyer [9] where the quantum player always beats the classic player in a *"mano a mano"* game. For a two women set Q uses only one step, but C needs two steps to find the right partner. In this case Q does better when $P_q > P_c/4$. Winning conditions improve for the quantum player for increasing N_w, but not in a monotonous way, because the number of steps used by Grover algorithm in Q search is an integer that increases in discrete steps.

In order to facilitate comprehension the set size in the simulations results shown is $N_w = 8$, that is the biggest N_w (taken as 2^n) in which Q uses only one step in Grover algorithm.

Under the first game conditions both players have only one attempt by turn. Since C cannot modify state ψ_1 amplitudes, he has $1/8$ chance to be right. On the other hand player Q, using Grover algorithm as his strategy, can modify states amplitudes in order to increase

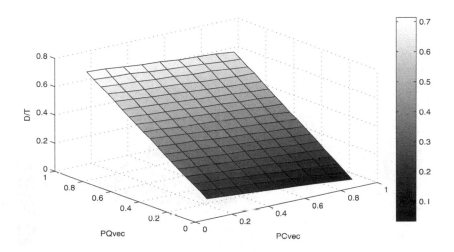

Figure 4. First game: One attempt for both players. Mean relative difference between Q and C success total number as $D/T = \frac{Qsuccess - Csuccess}{T}$, for different woman acceptation probabilities P_c^i and P_q^i. Q outperforms C in all shown cases. The small region where C prevails is not shown

his chances to win, reaching 0.78 as the probability to find his preferred woman in only one iteration. Figure 4 shows that situation outcomes for different P_c^i and P_q^i combinations. The vertical axis depicts D/T values as a function of P_c^i and P_q^i respectively. D/T is positive for all P_c^i and P_q^i values used in the simulation, which means that even at extremes where $P_c^i >> P_q^i$, the quantum player performs better. However there is a very small region where $P_c^i \approx 1$ and $P_q^i \approx 0$ not shown in the figure that corresponds to a prevailing C.

Under the second game conditions player C have $\frac{N_w}{2} = 4$ attempts before Q plays. After each C attempt the system is forced to collapse to one base state, so a third party, that could be the oracle, arrange the states again and mark the solution. As we explained above, to mark a state means to change its phase but nothing happens to the state amplitude, consequently, for the classic player C, the probability that state results the one the Oracle have signaled is, marked or not, $1/N_w = 1/8$, even though, due to his "insistence", he tries $\frac{N_w}{2} = 4$ times, his dating success chances increase considerably with respect to the first case. Figure 5 shows the corresponding results, where it is possible to see that classic player C begins to outperform Q when $P_c^i >> P_q^i$, that is, when woman has a marked preference for player C.

Player C probability to find the chosen woman can increase to $\frac{1}{2}$ when using a classical algorithm like "Brute-Force algorithm". As shown in figure 5, when C has $\frac{N_w}{2} = 4$ tries while Q has only one, C's odds of success in dating increases, and there are zones on the graph where $D/T < 0$. This implies that player C outperforms player Q. Nevertheless, to achieve that, the chosen woman preferences must be considerably greater for the classic player, that is $P_c^i > 2P_q^i$.

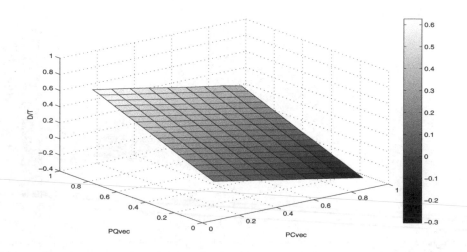

Figure 5. Second game: Classic player C has four tries while Q has only one. Mean relative difference between Q and C success total number as $D/T = \frac{Qsuccess - Csuccess}{T}$, for different woman acceptation probabilities P_c^i and P_q^i. C outperforms Q when $P_c^i \gg P_q^i$

4.1. Section discussion

In this section we have introduced a quantum formulation for decision matching problems, specifically for the dating game. In that framework women are represented with quantum states whose associated amplitudes must be modified by men's selection strategies, in order to increase a particular state amplitude and to decrease the others, with the final purpose to achieve the best possible choice when the game finishes. This is a highly time consuming task that takes a $O(N)$ runtime for a classical probabilistic algorithm, being N the women database size. Grover quantum search algorithm is used as a playing strategy that takes the man $O(\sqrt{N})$ runtime to find his chosen partner. As a consequence, if every man uses quantum strategy, no one does better than the others, and stability is quickly obtained.

The performances of quantum vs. classic players depend on the number of players N. In a "one on one" game there is no advantage from any of them and the woman preferences rule. Similar chances for quantum and classic players in "one on one" situation is not usual in most quantum games. Winning conditions improve for the quantum player for increasing N and the same number of attempts, but not in a monotonous way. The comparison between quantum and classic performances shows that for the same numbers of attempts, the quantum approach outperforms the classical approach. If the game is set in order that the classic player has $\frac{N}{2}$ opportunities and the quantum player only one, the former player begins to have an advantage over the quantum one when his probability to be accepted by the chosen woman is much higher than the probability for the quantum player.

5. Stability of couples

There is a group of N_m men and N_w women playing the game. Be $S_i = \{|0\rangle, |1\rangle, ..., |N_w - 1\rangle\}$ the states in a Hilbert space of man i decisions, where $\{0, 1, ..., N_w - 1\}$ are indexes in decimal notation identifying all the women he may choose. As a result each man has been assigned $log_2(N_w)$ qubits in order to identify each woman. Generally, the state vector of one man decisions will be in quantum superposition of the base states, $\Psi_i = \sum_{j=0}^{N_w-1} \alpha_j |j\rangle$, where $|\alpha_j|^2$ is the probability that man i selects woman j when system state is Ψ_i so must satisfied the normalization condition $\sum_{j=0}^{N_w-1} |\alpha_j|^2 = 1$. If there is no correlation between players, the state space of all men decision system is represented through $S_M = S_0 \otimes S_1 \otimes ... \otimes S_{M-2} \otimes S_{M-1}$, where $\otimes$ is the Kronecker product. Note that the S_M extends to any possible combination of men elections. On the other side there are the women who receive men proposals and must decide whether to accept or not one of them. With greater or lesser probability they will receive the all men's proposals, so following the same argument used with the men, be $\Psi_j = \sum_{i=0}^{M-1} \alpha_i |i\rangle$ the woman j acceptation state and be $S_W = S_0 \otimes S_1 \otimes ... \otimes S_{N_w-2} \otimes S_{N_w-1}$ the women acceptances space state. Finally, to close the circle, we define the couples possible states which must include so all possible men's elections as all possible women's acceptances. Accordingly, state space of the couples emerge from the Kronecker product of the men and women spaces, i.e. $S_C = S_M \otimes S_W$.

5.1. Strategies

In quantum games, players strategies are represented by unitary operators, which in quantum mechanics are also known as evolution operators related to the system's Hamiltonian [27]. If we call U_i the operator corresponding to player i strategy, the N-players strategies operator results $U = U_0 \otimes U_1 ... U_i \otimes ... \otimes U_{N-1}$. Starting from the initial pure state $|\Psi_0\rangle$ of the system, players apply their strategies U in order to modify it according to their preferences, that is modifying the probability amplitudes associated with each base state. As a consequence, evolution from the initial system state to some state $|\Psi_1\rangle$ is given by $|\Psi_1\rangle = U|\Psi_0\rangle$. Note that, following the reasoning of the preceding paragraph, when Ψ_0 is the initial state and Ψ_1 is the final state of the couples system, U arises from men and women strategies U_M and U_W respectively through $U = U_M \otimes U_W$. That is, U_M is applied by men to the qubits that identify the women states, meanwhile the women action on the qubits that identify men states is given by U_W.

5.2. Density matrix and system entropy

Often, as in life, players are not completely sure about which strategy to apply, that is, by the way of example, the case where someone chooses between the strategy U_a with probability p_a and U_b with probability $p_b = 1 - p_a$, that situation is referred in a mixed strategies game. Despite the complete system can be represented by its state vector, when it comes to mixed states the density matrix is more suitable. It was introduced by von Neumann to describe a mixed ensemble in which each member has assigned a probability of being in a determined state. The density operator, as it is also commonly called, represents the statistical mixture of all pure states and is defined by the equation

$$\rho = \sum_i p_i |\Psi_i\rangle\langle\Psi_i|, \tag{6}$$

where the coefficients p_i are non-negative and add up to one. From the density operator we can construct and understand the statistical behavior about our system by using statistical mechanics and a criterion of maximum or minimum entropy. Continuing the example, if it is supposed that the system starts in the pure state $\rho_0 = |\Psi_0\rangle\langle\Psi_0|$, after players mixed actions density matrix evolution is

$$\rho_1 = p_a U_a \rho_0 U_a^\dagger + p_b U_b \rho_0 U_b^\dagger. \tag{7}$$

Entropy is the central concept of information theories, [33]. The quantum analogue of entropy was introduced in quantum mechanics by von Neumann,[34] and it is defined by the formula

$$S(\rho) = -Tr\{\rho\log_2\rho\}. \tag{8}$$

5.3. $N = 2$ Model

In order to set up the notation let us look at the following example of two men and two women that interact for T times periods. Let define $\Psi_0^i = \alpha|0\rangle + \beta|1\rangle$ as the initial decision state of men i which is a linear superposition of the two possible options he has, they are woman 0 or 1. Without losing generalization consider $\alpha = 1$ and $\beta = 0$ which is consistent with thinking that they both have preference for the most popular, the most beautiful, the richest, or any superficial feature that most of the time makes men desire a woman at first glance. Consequently, the men's initial state vector is $\Psi_0^M = \Psi_0^0 \otimes \Psi_0^1 = |00\rangle$, where the first qubit represent man's 0 choice and the second is man's 1 choice. As we explain above, the initial quantum pure state is not stable, so during the game the state will change to the general form $\Psi_a^M = \sum_{i=0,j=0}^1 \alpha_{ij}|ij\rangle$ with probability p_a and $\Psi_b^M = \sum_{i=0,j=0}^1 \beta_{ij}|ij\rangle$ with probability p_b. As women have the last decision, they must evaluate men proposals and

decide to accept one of them or reject all. We consider, just for the example that woman 0 chooses man 0 with p_{0m} and man 1 with probability $1 - p_{0m}$, similar condition for woman 1 but in this case being p_{1m} the probability to choose man 0. That condition doesn't affect system stability but depending on the probabilities values does affect the maximum and minimum of the couple system's entropy. Equation 9 shows the women density matrix which has no off diagonal elements.

$$\rho_{w0} = p_{0m}p_{1m}|00\rangle\langle00| + p_{0m}(1-p_{1m})|01\rangle\langle01| + (1-p_{0m})p_{1m}|10\rangle\langle10| + (1-p_{0m})(1-p_{1m})|11\rangle\langle11|. \tag{9}$$

The direct product of all possible men proposals with all possible women decisions generates a possible partners state vector which in decimal notation is $\Psi_0^P = \sum_{i=0}^{15}|i\rangle$. Index i is a four qubits number, the first two qubits represent men 0 and 1 choices respectively and the

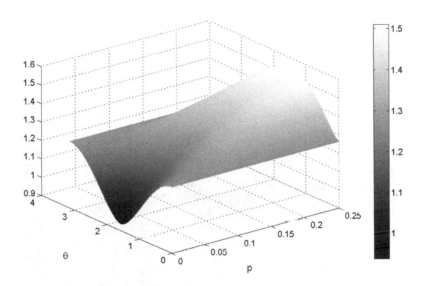

Figure 6. Quantum entropy corresponding to the situation where player 0 varies the probability p to apply strategy U_0^0

other two are the two women possible selections, then 16 are the possible couples states. For example, the state $|0101\rangle$ corresponds to the case that man 0 chooses woman 0 and she accepts him and the same occurs with man 1 and woman 1. Note that not all states corresponds to possible dates, some of them are considering the cases where there are no date, or the ones where only one couple is formed, the state $|0001\rangle$ is an example of the last case where the man man 0 chooses woman 0 and she accepts but on the other hand man 1 also chooses woman but she doesn't and woman 1 does not receive any proposition. As the game progress, probability amplitudes associated with mismatches must decrease, that because it is considered that people prefer to be coupled.

Single players moves or strategies are associated with unitary operators $U_i(\theta)$, with $0 \leq \theta \leq \pi$, applied on each one of their qubits, that in the general case where players have 2^n options, each pure strategy U is composed by n different $U_i(\theta_k)$, being k each state qubit. The general formula of U_i is 10, that are rotation operators, as explained in [27] any qubit operation can be decomposed as a product of rotations. In this work we consider $\gamma = 0$, therefore in what follows $U(\theta, 0)$ is always replaced by the simplest notation $U(\theta)$.

$$U(\theta, \gamma) = \begin{pmatrix} e^{i\gamma} \cdot \cos(\theta/2) & \sin(\theta/2) \\ -\sin(\theta/2) & e^{-i\gamma} \cdot \cos(\theta/2) \end{pmatrix} \tag{10}$$

Let p_0 be the probability of player 0 to apply strategy U_0^0 and $1 - p_0$ the probability to apply strategy U_1^0 to the initial state Ψ_0^i, while U_0^1 and U_1^1 are the strategies the man 1 applies with probability p_1 and $1 - p_1$ respectively. The strategies operators used in the examples are defined below, equations 11 and 12 are applied by man 0. Both of them transform the initial

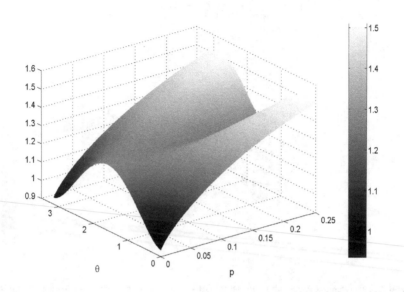

Figure 7. Quantum entropy corresponding to the situation where player 1 varies the probability p to apply strategy U_0^1

state $|0\rangle$ into states that are linear superpositions of 0 and 1, representing states with different probabilities of choosing one woman or the other. On the other hand, strategies applied by man 1 are presented in 13 and 14.

Figures 6 and 7 show two situations where the system entropy varies considerably as a function of the strategies the players use. Figure 2, for example, shows the case where the man 1 applies his strategies with fixed probability, just varying the angle θ while the other man (0) varies both strategies angle and the probability p. In all the cases we present here, in order to simplify the outcomes display, women density matrix doesn't change as explained above.

$$U_0^0 = U(\theta) \tag{11}$$
$$U_1^0 = U(\theta)U(\pi) \tag{12}$$

$$U_0^1 = U(\theta) \tag{13}$$
$$U_1^1 = U(-\theta) \tag{14}$$

For example, if both men choose the same woman with probability one, this is represented in Fig. 1 with $p = 0$. This situation is completely unstable because it is impossible for the woman to choose both of them at the same time (we assume). This correspond to minimum entropy as can be easily seen in Fig. 1. Depending on the strategies applied by men, the whole system entropy, that is the couple system entropy changes reaching maximum and minimum limits. As p increases, the mixing of the strategies also increases producing an increase in entropy that indicates a tendency to stability. The mixing of the strategies means that the men proposals are less focused on one woman. Fig. 2 shows the case where men's role change, that is man 0 fixes his strategies probabilities while 1 varies his own. Although for fixed θ angle, as expected, the minimum entropy points are located where player is applying a pure strategy ($p = 0$), for $\theta = \pi/2$ the entropy value does not change regardless of the value of p, because U_0^0 and U_1^0 are equivalent and therefore player 1 is applying a pure strategy. A result not shown in the figures is that entropy maxima increase when women preferences are the same for every men.

In this way, maxima and minima entropy points may be used to find stable states. Nevertheless, these stable states may not correspond to equilibria states of the game, because the players utilities has not been considered. In order to find Nash equilibria states, these utilities must be considered. This is beyond this chapter goals.

5.4. Section discussion

As a continuation of the analysis of a quantum formulation for the dating game that takes into account mutual progressive learning by representing women with quantum states whose associated amplitudes must be modified by men's selection strategies we concentrate on the information associated to the problem. Since we deal with mixed strategies, the density matrix formalism is used to describe the system. Even though the decisions players make are based on their payoffs, past experiences, believes, etc., we are not interested in that causes but in the consequences of the decision they take, that is, the influence of the strategies they apply on the quantum system stability by means of the equivalence between maximum entropy states and those states that obey the Collective Welfare Principle that says that a system is stable only if it maximizes the welfare of the collective above the welfare of the individual. Maxima and minima entropy points are used to find characteristic strategies that lead to stable and unstable states. Nevertheless, in order to find Nash equilibria states, the players utilities must be considered.

Maxima and minima entropy do not depend only on the strategies of men but also on women preferences, reaching the highest value when they have no preferences, that is when they choose every man with equal probability. On the other hand, minimum entropy correspond to men betting all chips to a single woman, without giving a chance to other woman.

6. Entangled strategies

The quantum dating market problem has been formulated as a two-sided bandit model [28], where in one side there are the men who must choose one "item" from the other side, which unlike the one side bandit, is composed by women able to reject the invitations.

The quantum formulation, which was presented in previous section, proceeds by assigning one basis of a Hilbert state space to each woman. As a consequence, if N_w is number of women playing, $S_i = \{|0\rangle, |1\rangle, ..., |N_w - 1\rangle\}$ are the states in the Hilbert space representing a man i decisions, therefore every man needs at least $log_2(N_w)$ qubits to identify each woman.

The state of man i decisions is a quantum superposition of the base states, $\Psi_i = \sum_{j=0}^{N_w-1} \alpha_j |j\rangle$, where $|\alpha_j|^2$ is the probability that man i selects woman j when system state is Ψ_i and $|\cdot\rangle$ is known as Dirac's notation. The normalization condition is $\sum_{j=0}^{N_w-1} |\alpha_j|^2 = 1$. On the other side of the market, women receive men proposals and must decide which is the best, accept it and reject the others. Thus, $\Psi_j = \sum_{i=0}^{N_m-1} \alpha_i |i\rangle$ is woman j acceptation state. Finally, combining proposals and acceptances the couples space which is the Kronecker product of all men's and all women's spaces is defined, i.e. $S_C = S_M \otimes S_W$.

Men decision states are separable when there is no connection among players, that is, for instance, no man has any emotional bond with some other that could condition their actions, thus all men decision state, ψ_M, is defined as $\psi_M = \otimes_{i=1}^{M-1} \psi_i$. The same reasoning corresponds to women states. On the other hand, if there is some relationship between two or more men, their actions are non-locally correlated, that is, their decisions are far from being independent. John Stuart Bell shown in 1966 that systems in entangled states exhibit correlations beyond those explainable by local "hidden" properties, or in other words, a non-local connection appears when two quantum particles are entangled, [35]. Therefore, we will study the case with correlation between agents by means of quantum entanglement, in other words, how harmful or beneficial can be for players knowing each other in advance.

As we mention in the previous section, players strategies are represented by unitary operators in quantum games. Starting the system in some state $|\Psi_0\rangle$ at time t_0, players apply their strategies U in order to modify it according to their preferences, that is modifying the probability amplitudes associated with each base state. Thus, evolution from the initial system state to some state $|\Psi_1\rangle$ in time t_1 is given by $|\Psi_1\rangle = U|\Psi_0\rangle$. The strategy operator U arises from men and women preferences operators U_M and U_W respectively through $U = U_M \otimes U_W$, where U_M is applied by men to the qubits that identify the women states, meanwhile the women action on the qubits that identify men states is given by U_W.

In order to understand the problem we analyze here a simple example of two men and two women. Single players moves or strategies are associated with 2×2 unitary rotation operators $U_i(\theta, \gamma)$ applied on each one of their qubits (15), where $0 \leq \theta \leq \pi$ and $0 \leq \gamma \leq \pi/2$. Men choices are coded by states $|w_0\rangle = |0\rangle$ and $|w_1\rangle = |1\rangle$, meanwhile women must decide between men $|m_0\rangle = |0\rangle$ and $|m_1\rangle = |1\rangle$. Since any qubit operation can be decomposed as a product of rotations, strategies combinations and possible outcomes are infinite. As a consequence, focusing on men relationship, we study three relevant cases. We suppose, as a measure of satisfaction, that men receive some payoff p_{w_i} if accepted by woman w_i, so for the example we have considered that $p_{w0} = 2$ and $p_{w1} = 5$.

$$U(\theta, \gamma) = \begin{pmatrix} e^{i\gamma}cos(\theta/2) & sin(\theta/2) \\ -sin(\theta/2) & e^{-i\gamma}cos(\theta/2) \end{pmatrix} \tag{15}$$

6.1. Results

For the first case, let us consider $\psi_0 = \frac{\sqrt{2}}{2}(|01\rangle + |10\rangle)$ as the initial state of men decisions system, where the left qubit of ψ_0 is representing man 0 election while the right one represents man 1 choice. As men states are entangled, it is not possible to uncouple their single actions. Therefore judging on the probability amplitudes, there is 50% probability that man 0 chooses woman 0 while man 1 chooses woman 1 and the other 50% for the other case. Since there is no way that men choose the same woman it is a state of mutual cooperation. Women acceptation state is initialized to $\psi_w = 0.5(|00\rangle + |01\rangle + |10\rangle + |11\rangle)$, implying that there is initially 25% chance that they choose the same man. In order to analyze the effect of woman behavior on men payoffs, for this and the following two cases, we consider that men decision state ψ_0 is invariant, while women strategies and their acceptation state ψ_w change. Figure 8 shows the payoff for man 0 as a function of women strategies which are set as $U_i = (t \cdot \pi, t \cdot \pi/2)$ for $t \in [-1, 1]$ and $i = 0, 1$. Different strategies imply changes on women preferences, so some change in $U_i(\theta, \gamma)$ implies that woman i acceptation probability distribution is modified. Following [9], equation 16 represents man 0 payoff, where P_{00} and P_{01} are his chances to be accepted for a date with woman 0 and 1 respectively.

$$\$_{m0} = 2 \cdot P_{00} + 5 \cdot P_{01} \tag{16}$$

In the second example we introduce competition between men. The initial men state is given by $\psi_0 = \frac{\sqrt{2}}{2}(|00\rangle + |11\rangle)$. Figure 9 depicts again the resulting payoffs for man 0 as a function of women strategies. Finally a third case is considered where men decision state is $\psi_m = 0.5(|00\rangle + |01\rangle + |10\rangle + |11\rangle)$. In this case men make independent choices choosing one of four possible options with equal probability. Figure 10 show the resulting payoffs as a function of woman strategies.

As the figures show, the different scenarios present significant differences on payoff topology and maximum payoff values.

The cooperative situation presents the highest payoff compared with the competitive and the independent ones as shown in figure 8. Figures 9 and 10 show that a better payoff may be obtained in the competitive setup compared to the independent one, but on the other hand, also a much lower payoff for other women strategies may be available. The independent decision scenario is thus characterized by lowest maxima and less variation on payoffs.

6.2. Section discussion

We have considered the dating market decision problem under the quantum mechanics point of view with the addition of entanglement between players states. Women and men are represented with quantum states and strategies are represented by means of unitary operator on a complex Hilbert space. Men final payoff, considering payoff as a measure of satisfaction, depends on the woman he is paired with. If men decision states are entangled, their actions are non-locally correlated modeling competition or cooperation scenarios. Three examples are shown in order to illustrate the more usual scenarios. In two of them the men strategies are correlated in a cooperative and a competitive way respectively. In the other example men strategies are independent. Although cooperative and competitive strategies can drive to higher payoffs, changing of women preferences on those scenarios can lead to very low payoffs. The independent decision scenario is characterized by less variation on payoffs.

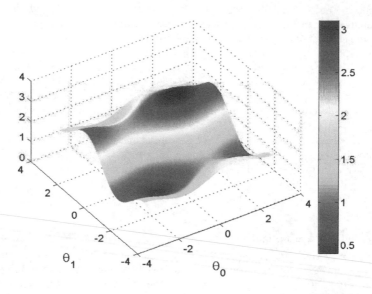

Figure 8. Payoff for man 0 if men never choose the same woman, as function of women acceptation strategies. For the example γ varies as $\theta/2$.

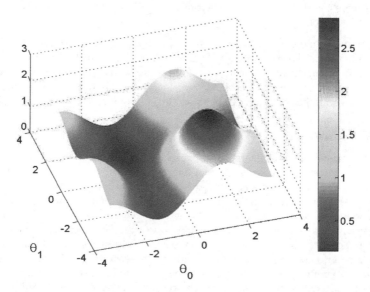

Figure 9. Payoff for man 0 if men always choose the same woman, as function of women acceptation strategies. For the example γ varies as $\theta/2$.

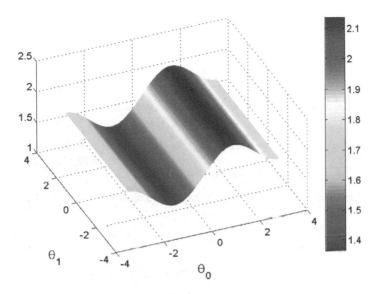

Figure 10. Payoff for man Ω if men choose without restrictions, states are not entangled, as function of women acceptation strategies. For the example γ varies as $\theta/2$.

7. Final remarks

The dating market problem may be included in a more general category of matching problems where the elements of two sets have to be matched by pairs. Matching problems have broad implications not only in economic and social contexts but in other very different research fields such as communications engineering or molecular biology, for example. The main goal of this chapter is to introduce and analyze a quantum formulation for the dating market game, whose nearest classical antecedent is the Stable Marriage Problem. Players strategies are represented by unitary operators, which in quantum mechanics are also known as evolution operators related to the Hamiltonian of the system. Significant outcomes arise when classic players play against quantum ones. For instance, when a quantum player uses Grover search algorithm as her strategy, her winning probabilities grow with increasing number of players, but none leads in a "one on one" game. Besides, from stability point of view, maxima and minima entropy points are used to find characteristic strategies that lead to unstable and stable states, resulting the highest entropy values when women have no preferences, that is, when they choose every man with equal probability. On the other hand, minimum entropy correspond to men betting all chips to a single woman, without giving a chance to other woman. Finally, to model relationships between people that may correlate their decisions, our model consider the situation when men decision states are entangled and their actions are non-locally correlated modeling competition or cooperation scenarios. One of the main outcomes is for example that, although cooperative and competitive strategies can drive to higher payoffs, changing of women preferences on those scenarios can lead to very low payoffs.

Author details

C. M. Arizmendi and O. G. Zabaleta

Facultad de Ingeniería, Universidad Nacional de Mar del Plata, Argentina

References

[1] P.I. Hagouel and I.G. Karafyllidis. Quantum computers: Registers, gates and algorithms. In *Microelectronics (MIEL), 2012 28th International Conference on*, pages 15 –21, may 2012.

[2] D.C. Marinescu. The promise of quantum computing and quantum information theory -quantum parallelism. In *Parallel and Distributed Processing Symposium, 2005. Proceedings. 19th IEEE International*, page 112, april 2005.

[3] H. Haffner, C.F. Roos, and R. Blatt. Quantum computing with trapped ions. *Physics Reports*, 469(4):155–203, 2008.

[4] Chao-Yang Pang, Ri-Gui Zhou, Cong-Bao Ding, and Ben-Qiong Hu. Quantum search algorithm for set operation. *Quantum Information Processing*, pages 1–12, 2012.

[5] B. Criger, O. Moussa, and R. Laflamme. Quantum error correction with mixed ancilla qubits. *Phys Rev A*, 85(4):5 pp, 2012.

[6] P. W. Shor. Polynomial-time algorithms for prime factorization and discrete logarithms on a quantum computer. *arXiv:quant-ph/9508027v2*, 1996.

[7] M. Boyer, G. Brassard, P. Hoeyer, and Tapp A. Tight bounds on quantum searching. *Fortsch.Phys*, 46:493–506, 1998.

[8] L.K. Grover. A fast quantum mechanical algorithm for database search. In *Proceedings, 28th Annual ACM Symposium on the Theory of Computing (STOC)*, pages 212–219, May 1996.

[9] D.A. Meyer. Quantum strategies. *Phys. Rev. Lett.*, 82:1052–1055, 1999.

[10] J. Eisert, M. Wilkens, and M. Lewenstein. Quantum games and quantum strategies. *Phys. Rev. Lett.*, 83:3077–3080, 1999.

[11] A. Lambert-Mogiliansky, S. Zamir, and H. Zwirn. Type indeterminancy: a model of kt-man. *J. Math. Psych.*, 53:349–361, 2009.

[12] V.I. Yukalov and D. Sornette. Quantum decision theory as quantum theory of measurement. *Physics Letters A*, 372(46):6867–6871, 2008.

[13] T. Temzelides. An uncertainty principle for social science experiments. *Available at http://www.owlnet.rice.edu/ tl5*, 2005.

[14] C. M. Arizmendi and O. G. Zabaleta. Stability of couples in a quantum dating market. *special IJAMAS issue "Statistical Chaos and Complexity*, 26:143–149, 2012.

[15] O.G. Zabaleta and C.M Arizmendi. Quantum decision theory on a dating market. *Advances and Applications in Statistical Sciences (AASS)*, 6:489, 2011.

[16] D. Gale and L.S. Shapley. College admissions and the stability of marriage. *Am. Math. Monthly*, 69:9–15, 1962.

[17] Y. Zhang and J Leezer. Simulating human-like decisions in a memory-based agent model. *Computational and Mathematical Organization Theory*, 16(4):373–399, 2009.

[18] P. Laureti and Y.C. Zhang. Matching games with partial information. *Physica A*, 324:49–65, 2003.

[19] A.E. Roth and M. Sotomayor. Two-sided matching: A study in game-theoretic modeling and analysis. *Econometric Society Monograph Series*, 324, 1990.

[20] R. V. Mendes. The quantum ultimatum game. *Quantum Information Processing*, 4(1):1–12, 2005.

[21] K.B. Clark. Origins of learned reciprocity in solitary ciliates searching grouped 'courting' assurances at quantum efficiencies. *Biosystems*, 99:27–41, 2010.

[22] O.G. Zabaleta and C.M Arizmendi. Quantum dating market. *Physica A*, 389:2858–2863, 2010.

[23] Esteban Guevara Hidalgo. Quantum games entropy. *Physica A*, 383:797–804, 2007.

[24] B Arfi. Quantum social game theory. *Physica A*, 374:794–820, 2007.

[25] M. J. Osborne and A. Rubinstein. *A Course in Game Theory*. The MIT Press Cambridge, Massachusetts, England, 1998.

[26] Staphane Vialette. On the computational complexity of 2-interval pattern matching problems. *Theoretical Computer Science*, 312(23):223 – 249, 2004.

[27] M.A. Nielsen and I.L. Chuang. *Quantum Computation and Quantum Information*. Cambridge University Press, England, 2000.

[28] S. Das and E. Kamenica. Two-sided bandits and the dating market. In Alessandro Saffiotti Leslie Pack Kaelbling, editor, *Proceedings of the Nineteenth International Joint Conference on Artificial Intelligence,IJCAI-2005*, pages 947–952, Edinburgh, Scotland, UK, August 2005.

[29] C.M Arizmendi. Paradoxical way for losers in a dating game. In *Proc. AIP Nonequilibrium Statistical Mechanics and Nonlinear Physics*, pages 20–25, Mar del Plata , Argentina, December 2006.

[30] A. Romanelli. Quantum games via search algorithms. *Physica A*, 379:545–551, 2007.

[31] Chao-Yang Pang, Cong-Bao Ding, and Ben-Qiong Hu. Quantum pattern recognition of classical signal. *quant-ph-arXiv:0707.0936v2*, 2007.

[32] M.-J. Omero, M. Dzierzawa, M. Marsili, and Y.-C. Zhang. *J. Physique*, 7:1723, 1997.

[33] C Shannon. A mathematical theory of communication. *Bell System Tech. Jour.*, 27:379–423, 1948.

[34] J. von Neumann.   *Mathematische Grundlagen der Quantenmechanik.*   Springer-Verlag [translated by R. T. Beyer as Mathematical Foundations of Quantum Mechanics (Princeton University Press, Princeton, 1955)], Berlin, 1932.

[35] James D. F. V. and P. G. Kwiat. Quantum state entanglement. *Los alamos Science*, 27:52–57, 2002.

Chapter 2

Quantization as Selection
Rather than Eigenvalue Problem

Peter Enders

Additional information is available at the end of the chapter

1. Introduction

The experimental, in particular, spectroscopic results about atoms brought Bohr [3] to the following "principal assumptions" for a theory of atoms.

1. That the dynamical equilibrium of the systems in the stationary states can be discussed by help of the ordinary mechanics, while the passing of the systems between different stationary states cannot be treated on that basis.

2. That the latter process is followed by the emission of a homogeneous radiation, for which the relation between the frequency and the amount of energy emitted is the one given by Planck's theory." (p. 7)

Assumption (1) states a strange contraposition of the conservation and of the changes of stationary states. Indeed, the conservation of energy - a generalization of Newton's Law 1 - holds true quite general, while the change of stationary states along Newton's Law 2 is restricted to classical mechanics (CM). A smooth transition from representations of CM, which axiomatically fix not only the conditions of conservation, but also the manner of change of stationary states and the equation(s) of motion, respectively (Newton, Lagrange, Hamilton), is impossible, as observed also by Heisenberg [30] and Schrödinger [43]. However, in Leonhard Euler's representation of CM [24][25], only the conditions of conservation of stationary states are fixed, while their change has to be described according to the situation under consideration. This makes it suitable to serve as starting point for a smooth transition from classical to quantum mechanics (QM). Euler's principles of the change of stationary states of bodies will be generalized to classical conservative systems as well as to quantum systems. The latter will be used for deriving the time-dependent from the time-independent Schrödinger equation.

In his pioneering papers 'Quantization as Eigenvalue Problem' [43], Schrödinger has posed four requirements.

1. The "quantum equation" should "carry the quantum conditions in itself" (Second Commun., p. 511);

2. There should be a special mathematical method for solving the stationary Schrödinger equation, which accounts for the *non*-classical character of the quantization problem, *ie*, which is different from the classical eigenvalue methods for calculating the (eigen)modes of strings, resonators and so on (*ibid.*, p. 513);

3. The derivation should uniquely decide, that the energy rather than the frequency values are discretized (*ibid.*, pp. 511, 519), since only the former means true quantization, while the latter corresponds to the classical discretization mentioned in requirement 2;

4. The use of the classical expressions for the kinetic and potential energies should be justified (Fourth Commun., p. 113).

Schrödinger's requirements 1 and 2 mean, that - contrary to the very title of the papers - quantization is actually *not* an eigenvalue problem. For in the latter the discretization is imposed not by the differential equation itself, but by the boundary conditions, and this is the classical discretization for standing waves in organ pipes etc. I will fulfill all four requirements by treating quantization as a selection problem. The number of stationary states of a quantum system is smaller than that of a classical system [6]. I will describe, (i), the selection of quantum systems out of the set of all mechanical systems and, (ii), the selection of their stationary states out of the continuum of classical stationary states. Earlier arguing [19][10] is improved and extended in several essential points.

2. Elements of an Eulerian representation of classical mechanics

2.1. Euler's axioms

Leonhard Euler [20-25] was the first to apply the calculus to all areas of mathematics and mechanics of his time, and he developed novel methods, such as the calculus of variations and topology. Moreover, he worked out an axiomatic of mechanics, where only Newton's Law (axiom) 1 concerning the conservation of stationary states is retained as an axiom, while Newton's Laws (axioms) 2 and 3 concerning the change of stationary states are treated as problems to be solved (for a detailed account, see [19][10][45]). This allows for introducing alternative equations of motion without loosing the contact to CM.

The existence of stationary states is postulated in the following axioms (as in Newton's axioms, rotatory motion is not considered).

Axiom E0: Every body is either resting or moving.

This means, that the subsequent axioms E1 and E2 are not independent of each other; they exclude each other and, at once, they are in harmony with each other [22].

Axiom E1: A body preserves its stationary state at rest, unless an external cause sets it in motion.

Axiom E2: A body preserves its stationary state of straight uniform motion, unless an external cause changes this state.

The stationary-state variable is the velocity vector, $\mathbf{v}$ (the mass of a body is always constant). The equation of stationary state reads $\mathbf{v}=0$ for the state at rest and $\mathbf{v}=const$ for the state of straight uniform motion. The position, $\mathbf{r}$, is variable of the state of motion, but not of stationary states, because it changes during straight uniform motion, *ie*, in the absence of (external) causes for changing the stationary state [50].

2.2. Eulerian principles of change of state for single bodies

Following [21], the changes of position, $\mathbf{r}$, and velocity, $\mathbf{v}$, of a body of mass m subject to the (external) force, $\mathbf{F}$, during the time interval dt are

$$d\binom{\mathbf{r}}{\mathbf{v}}=\binom{\mathbf{v}dt}{\frac{1}{m}\mathbf{F}dt}=\begin{pmatrix}0&1\\0&0\end{pmatrix}\binom{\mathbf{r}}{\mathbf{v}}dt+\begin{pmatrix}0&0\\0&\frac{1}{m}1\end{pmatrix}\binom{0}{\mathbf{F}}dt\equiv\mathbf{U}_{int}\binom{\mathbf{r}}{\mathbf{v}}+\mathbf{U}_{ext}\binom{0}{\mathbf{F}} \tag{1}$$

The internal transformation, $\mathbf{U}_{int}$, describes the internal change, $d\mathbf{r}=\mathbf{v}dt$, that is independent of the external force. The external transformation, $\mathbf{U}_{ext}$, describes the external change, $d\mathbf{v}=(\mathbf{F}/m)dt$, that depends on the external force. These matrices do not commute: $\mathbf{U}_{int}\mathbf{U}_{ext}\neq\mathbf{U}_{ext}\mathbf{U}_{int}$. This means, that the internal and external transformations are not reducible onto another; the internal and external changes are independent each of another.

Thus, up to order dt,

CB1) The changes of stationary-state quantities ($d\mathbf{v}$) are external changes; they explicitly depend solely on external causes ($\mathbf{F}$), but not on state-of-motion quantities ($\mathbf{r}$);

CB2) The change of the stationary-state quantities ($d\mathbf{v}$) do not explicitly depend on the stationary-state quantities ($\mathbf{v}$) themselves;

CB3) The change of state-of-motion quantities ($d\mathbf{r}$) are internal changes; they explicitly depend solely on stationary-state quantities ($\mathbf{v}$) ;

CB4) The change of stationary-state ($d\mathbf{v}$) and of state-of-motion quantities ($d\mathbf{r}$) are independent each of another;

CB5) As soon as the external causes ($\mathbf{F}$) vanish, the body remains in the stationary state assumed at this moment: $Z(t)=const=Z(t_1)=\mathbf{v}(t_1)$ for $t\geq t_1$, if $\mathbf{F}(t)=0$ for $t\geq t_1$.

Accounting for $ddt=0$ and $d\mathbf{F}=0$, one obtains from eq. (1)

$$dd\begin{pmatrix} \mathbf{r} \\ \mathbf{v} \end{pmatrix} = \begin{pmatrix} 0 & 1 \\ 0 & 0 \end{pmatrix} d\begin{pmatrix} \mathbf{r} \\ \mathbf{v} \end{pmatrix} dt = \begin{pmatrix} 0 & 1 \\ 0 & 0 \end{pmatrix} \begin{pmatrix} 0 & 0 \\ 0 & \frac{1}{m}1 \end{pmatrix} \begin{pmatrix} 0 \\ \mathbf{F} \end{pmatrix} dt^2 = \begin{pmatrix} \frac{1}{m}\mathbf{F} \\ 0 \end{pmatrix} dt^2 \tag{2}$$

Thus, the principles CB1...CB5 are compatible with Newton's equation of motion (published first in [20]). For their relationship to Descartes' and Huygens' principles of motion, see [10][45].

2.3. Eulerian principles of change of state for Hamiltonian systems

According to Definition 2 and the axioms, or laws of motion (Laws 1 and 2, Corollary 3), the momentum is the stationary-state variable of a body in Newton's *Principia*. The total momentum "is not changed by the action of bodies on one another" (Corollary 3). The principles CB1...CB5 remain true, if the velocity, $\mathbf{v}$, is replaced with the momentum, $\mathbf{p}$. For this, I will use $\mathbf{p}$ rather than $\mathbf{v}$ in what follows.

For a free body, any function of the momentum, $Z(\mathbf{p})$, is a conserved quantity. If a body is subject to an external force, its momentum is no longer conserved, but becomes a state-of-motion variable like its position. Correspondingly, $Z(\mathbf{p}) \neq const$. Suppose, that there is nevertheless a function, $Z_0(\mathbf{p},\mathbf{r})$, that is constant during the motion of the body. It describes the stationary states of the system body & force. External influences (additional forces) be described through a function $Z_{ext}(\mathbf{p},\mathbf{r},t)$ such, that $Z(\mathbf{p},\mathbf{r})=Z_0(\mathbf{p},\mathbf{r})+Z_{ext}(\mathbf{p},\mathbf{r},t)$ takes over the role of the stationary-state function.

The following principles - a generalization of CB1...CB5 - will shown to be compatible with Hamilton's equations of motion. Up to order dt,

CS1) The change of stationary-state quantities (dZ) depends solely on the external influences (Z_{ext}), but not on state-of-motion quantities ($\mathbf{p}$, $\mathbf{r}$);

CS2) The change of the stationary-state quantities (dZ) is independent of the stationary-state quantities (Z) themselves;

CS3) The changes of state-of-motion quantities ($d\mathbf{p}$, $d\mathbf{r}$) directly depend solely on stationary-state quantities (Z); the external influences (Z_{ext}) affect the state-of-motion quantities ($\mathbf{p}$, $\mathbf{r}$) solely indirectly (via stationary-state quantities, Z);

CS4) The changes of stationary-state (dZ) and of state-of-motion quantities ($d\mathbf{p}$, $d\mathbf{r}$) are independent each of another;

CS5) As soon as the external influences (Z_{ext}) vanish, the system remains in the stationary state assumed at this moment: $Z(t)=const=Z(t_1)$ for $t \geq t_1$, if $Z_{ext}=0$ for $t>t_1$.

These principles imply the equation of change of stationary state to read

$$dZ = \frac{\partial Z}{\partial \mathbf{p}} \cdot d\mathbf{p} + \frac{\partial Z}{\partial \mathbf{r}} \cdot d\mathbf{r} + \frac{\partial Z}{\partial t} dt = \frac{\partial Z_{ext}}{\partial t} dt$$

This equation is fulfilled, if

$$\frac{\partial Z}{\partial \mathbf{p}} = a\frac{d\mathbf{r}}{dt}; \qquad \frac{\partial Z}{\partial \mathbf{r}} = -a\frac{d\mathbf{p}}{dt} \tag{4}$$

Compatibility with Newton's equation of motion yields $a=1$ and $Z(\mathbf{p},\mathbf{r})=H(\mathbf{p},\mathbf{r})$, the Hamiltonian of the system; (4) becoming Hamilton's equations of motion.

$$\frac{d\mathbf{p}}{dt} = -\frac{\partial H}{\partial \mathbf{r}}; \qquad \frac{d\mathbf{r}}{dt} = \frac{\partial H}{\partial \mathbf{p}} \tag{5}$$

It may thus be not too surprising that these principles can *cum grano salis* be applied also to quantum systems. Of course, the variables, which represent of stationary states and motion, will be other ones, again.

3. Quantization as selection problem — I. Derivation of the stationary Schrödinger equation

The usual foundations of QM consider CM to be not sufficient and, consequently, need additional or novel assumptions, for instance,

- to restrict the energy spectrum to the values $nh\nu$ [41][6] or to $(n/2)h\nu$ [3];

- to "distinguish" [31] or to "select" [39][36] the values nh of the action integral, $\oint p\,dq$ (n - integer; in contrast to CM, the action integral is *not* subject to a variational principle);

- to suppose the existence of h and to abandon the classical paths [30];

- to suppose the existence of h and of a wave function being the solution of an eigenvalue problem [43];

- to suppose the existence of a quantum logic [2][28] and of a Hilbert space for its representation;

- to suppose the existence of transition probabilities obeying the Chapman-Kolmogorov equation (as in wave mechanics) [27].

All these approaches have eventually resorted to CM in using the classical expressions *and* the interpretations of position, momentum, potential and kinetic energies, because 'it works'. In contrast, I will present a concrete realization of Schrödinger's 4th requirement.

3.1. The relationship between CM and non-CM as selection problem

In his Nobel Award speech, Schrödinger ([44] p. 315) pointed to the logical aspect, which is central to the approach exposed here.

"We are faced here with the full force of the logical opposition between an

either – or (point mechanics)

and a

both – and (wave mechanics)

This would not matter much, if the old system were to be dropped entirely and to be *replaced* by the new."

This "logical opposition" consists in a hierarchy of selection problems.[1]

3.1.1. Selection problem between Newtonian and non-Newtionian CM

Consider a linear undamped oscillator. For each stationary state of total energy E, Newton's equation of motion confines its position, x, to the interval between and including the classical turning points: $x_{min} \leq x \leq x_{max}$. Its momentum is confined as $p_{min} \leq p \leq p_{max}$. More generally speaking, a system obeying Newton's equation of motion moves within the sets $C^{Newton}=\{\mathbf{r} \mid V(\mathbf{r}) \leq E\}$ and $P^{Newton}=\{\mathbf{p} \mid T(\mathbf{p}) \leq E\}$.[2]

Alternatively, a classical, though non-Newtonian mechanics is conceivable, where $dp/dt=-\mathbf{F}$ and $E=V-T$. A linear oscillator would move beyond the turning points: $x \leq x_{min}$ or/and $x \geq x_{max}$. In general, the set of possible configurations equals $C^{non-Newton}=\{\mathbf{r} \mid V(\mathbf{r}) \geq E\}$. The momentum configuration is no longer limited: $P^{non-Newton}=P^{all}=\{\mathbf{p}\}$.

Thus, a conservative CM system obeys either the laws of Newtonian CM, where $dp/dt=+\mathbf{F}$ and $V(\mathbf{r}(t)) \leq E$, or the laws of non-Newtonian CM, where $dp/dt=-\mathbf{F}$ and $V(\mathbf{r}(t)) \geq E$. In both cases, the system moves along paths, $\mathbf{r}(t)$.

3.1.2. Einsteinian selection problem between CM and non-CM

For both Newtonian and non-Newtonian classical systems, the set of possible energies (the energy spectrum) is continuous. Einstein [6] has observed that this leads to a temperature-independent specific heat (Dulong-Petit's law) for an ensemble of classical oscillators. In contrast, the discrete set of possible energies of a Planck oscillator yields a specific heat that decreases with decreasing temperature, in agreement with then recent experiments. He

---

1 I will deviate from the exposition in [19][10] to make it shorter, though clearer and to correct few statements about the momentum configurations.

2 $V(\mathbf{r}) \geq 0$, since it equals the "disposable work storage of a system" [33]. '$\mathbf{r}$' stays for all configuration variables, '$\mathbf{p}$' stays for all momentum configuration variables of a system.

concluded, "that the set of possible energies of microscopic systems is smaller than that for systems of our everyday experience."

Thus, the set of possible energies of a mechanical system is either continuous, or discrete.[3]

3.1.3. Selection problem between CM and non-CM in terms of allowed configurations

Einstein's alternative does not follow from purely mechanical arguing. For this, I continue the reasoning of subsection 3.1.1. The harmony between Newtonian and non-Newtonian CMs consists in that they both build an alternative to a non-CM mechanics, in which the set of allowed configurations comprises the *whole* configuration space, $C^{all} = C^{Newton} \cup C^{non-Newton} = \{r\}$. Since the motions along paths in C^{Newton} and $C^{non-Newton}$ are incompatible each to another, the motion of non-CM systems does not proceed along paths.

Thus, a mechanical system either moves along paths (CM), or it moves not along paths (non-CM). The configuration of a non-CM system can assume any element of C^{all} even in the stationary states.

3.1.4. Selection problem between mechanics and non-mechanics

For completeness I note that a system is either a mechanical, or a non-mechanical one.

Like Euler's axioms E1 and E2, these alternatives exclude each other and, at once, are "in harmony each with another". They dialectically determine each another in the sense of Hegel [29].

The question thus is how a linear oscillator *without turning points* is to be described?

3.2. Non-classical representation of the potential and kinetic energies

$V(r)$ [$T(p)$] is no longer the contribution of the (momentum) configuration r (p) to the total energy, E, since it is unbounded in the domain C^{all} (P^{all}). For this, I define 'limiting factors', $F_E(r)$ and $G_E(p)$, such, that

$$V_E^{ncl}(r) = F_E(r)V(r) \leq E; \qquad r \in C^{all}$$
$$T_E^{ncl}(p) = G_E(p)T(p) \leq E; \qquad p \in P^{all} \tag{6}$$

The contribution of the (momentum) configuration r (p) of a non-classical system to its total energy, $V_E^{ncl}(r)$ [$T_E^{ncl}(p)$], depends on the energy, because the inequality is no longer realized through the restriction of the (momentum) configuration space.

$F_E(r)$ and $G_E(p)$ are non-negative. $F_E(r)<0$ would mean, that $V_E^{ncl}(r)$ is attractive (repulsive), while $V(r)$ is repulsive (attractive). $G_E(p)<0$ would mean, that $T_E^{ncl}(p)$ becomes negative. For simplicity, I chose the one-dimensional representation of unity and set

---

3 Because of the finite resolution of measurement apparatus, the set of rational numbers is physically equivalent to the set of real numbers.

$$F_E(\mathbf{r}) = \left|f_E(\mathbf{r})\right|^2 ; \qquad G_E(\mathbf{p}) = \left|g_E(\mathbf{p})\right|^2 \tag{7}$$

"$\psi\bar{\psi}$ [$\equiv|\psi|^2$] is a kind of *weight function* in the configuration space of the system. The *wave-mechanical* configuration of the system is a *superposition* of many, strictly speaking, of *all* kinematically possible point-mechanical configurations. Thereby, each point-mechanical configuration contributes with a certain *weight* to the true wave-mechanical configuration, the weight of which is just given through $\psi\bar{\psi}$. If one likes paradoxes, one can say, the system resides quasi in all kinematically thinkable positions at the same time, though not 'equally strongly'. " ([43] 4th Commun., p. 135)

Correspondingly, I call F_E and G_E weight functions, f_E and g_E - weight amplitudes. Since $F_E(\mathbf{r})$ and $G_E(\mathbf{p})$ are dimensionless, there are reference values, r_{ref} and p_{ref}, such, that actually, $F_E(\mathbf{r})=F_E(\mathbf{r}/r_{ref})$ and $G_E(\mathbf{p})=G_E(\mathbf{p}/p_{ref})$. In other words, each such system has got a characteristic length in configuration and in momentum configuration space.

$$\iiint\limits_{C^{all}} F_E(\frac{\mathbf{r}}{r_{ref}})\frac{d^3r}{r_{ref}^3} = \iiint\limits_{p^{all}} G_E(\frac{\mathbf{p}}{p_{ref}})\frac{d^3p}{p_{ref}^3} = 1 \tag{8}$$

In order to simplify the notation, I will omit r_{ref} and p_{ref} wherever possible.

3.3. The stationary Schrödinger equation

Within CM, the balance between potential, $V(\mathbf{r})$, and kinetic energies, $T(\mathbf{p})$, to yield $V(\mathbf{r})$ $+T(\mathbf{p})=E=const$ is realized through the common path parameter time, t: $\mathbf{r}=\mathbf{r}(t)$, $\mathbf{p}=\mathbf{p}(t)$; $E=V(\mathbf{r}(t))$ $+T(\mathbf{p}(t))$. This common parameterization through t is absent for non-classical systems not moving along paths, $\mathbf{r}(t)$. Consequently, the balance between the potential, $V_E^{ncl}(\mathbf{r})$, and kinetic energies, $T_E^{ncl}(\mathbf{p})$, is not point-wise: $\mathbf{p}(t)\leftrightarrow\mathbf{r}(t)$, but set-wise: $\{\mathbf{p}\}\leftrightarrow\{\mathbf{r}\}$. Set-wise relations are mediated through integral relations.[4]

$$g_E(\mathbf{p}) = \frac{1}{\left(2\pi p_{ref}r_{ref}\right)^{\frac{3}{2}}}\iiint\limits_{C^{all}} e^{i\frac{\mathbf{p}\cdot\mathbf{r}}{p_{ref}r_{ref}}} f_E(\mathbf{r})d^3r; \qquad f_E(\mathbf{r}) = \frac{1}{\left(2\pi p_{ref}r_{ref}\right)^{\frac{3}{2}}}\iiint\limits_{p^{all}} e^{-i\frac{\mathbf{r}\cdot\mathbf{p}}{p_{ref}r_{ref}}} g_E(\mathbf{p})d^3p \tag{9}$$

In view of the symmetric normalization (8) I have chosen symmetric normalization factors.

Alternatively, it is possible to avoid complex-valued weight amplitudes (wave functions) through using 2-component vectors for them and the Hartley transform in place of the Fourier

---

4 The most general symmetric Fourier transform contains a free complex-valued parameter [49]. It appears to be merely a rescaling of r_{ref} and p_{ref}, respectively.

transform. The operators become 2x2 matrices. It remains to explore whether their free components can be exploited for the description of new effects.

Lacking orbits, such a system does not assume a definite configuration, say, r_1, and momentum configuration, p_1, at a given time, t_1, with $E=V_E^{ncl}(r_1)+T_E^{ncl}(p_1)$. Instead, they *all* contribute to the stationary state, E. The partial contribution of the single (momentum) configuration, r (p), is determined by the weight function according to eq. (6). The total energy thus becomes

$$E = \frac{\iiint\limits_{C^{all}} F_E(\mathbf{r})V(\mathbf{r})d^3r}{\iiint\limits_{C^{all}} F_E(\mathbf{r})d^3r} + \frac{\iiint\limits_{p^{all}} G_E(\mathbf{p})T(\mathbf{p})d^3p}{\iiint\limits_{p^{all}} G_E(\mathbf{p})d^3p} \tag{10}$$

The denominators have been added for dimensional reasons. The classical representation is obtained through setting

$$F_E(\mathbf{r}) = r_{ref}^3\delta(\mathbf{r}-\mathbf{r}(t)); \qquad G_E(\mathbf{p}) = p_{ref}^3\delta(\mathbf{p}-\mathbf{p}(t)) \tag{11}$$

The occurrence of E on the r.h.s. makes eq. (10) to be an implicit equation for E. This suggests E to be an *internal* system parameter being determined solely by system properties like the oscillation frequency of an undamped harmonic oscillator [19]. However, as in CM, the value of E is given by the initial preparation of the system.

The Fourier transform (9) enables me to eliminate one of the weight amplitudes from eq. (10).

$$E = \frac{\iiint\limits_{C^{all}} \bar{f}_E(\mathbf{r})\hat{H}(\mathbf{r})f_E(\mathbf{r})d^3r}{\iiint\limits_{C^{all}} \bar{f}_E(\mathbf{r})f_E(\mathbf{r})d^3r}; \qquad \hat{H}(\mathbf{r}) \equiv V(\mathbf{r})+T(-ip_{ref}r_{ref}\frac{\partial}{\partial\mathbf{r}})$$

$$= \frac{\iiint\limits_{p^{all}} \bar{g}_E(\mathbf{p})\hat{H}(\mathbf{p})g_E(\mathbf{p})d^3p}{\iiint\limits_{p^{all}} \bar{g}_E(\mathbf{p})g_E(\mathbf{p})d^3p}; \qquad \hat{H}(\mathbf{p}) \equiv V(ir_{ref}p_{ref}\frac{\partial}{\partial\mathbf{p}})+T(\mathbf{p}) \tag{12}$$

(Other positions of the weight amplitudes lead to the same results.) Since, in general, $\bar{f}_E$ and $\bar{g}_E$ are linearly independent of f_E and g_E, respectively, necessary conditions for fulfilling these equations are

$$Ef_E(\mathbf{r}) = \hat{H}(\mathbf{r})f_E(\mathbf{r}); \qquad Eg_E(\mathbf{p}) = \hat{H}(\mathbf{p})g_E(\mathbf{p}) \tag{13}$$

Moreover, these equations hold true for the minimum of the r.h.s. of eq. (12), *ie*, for the ground state. There is no indication for a difference between the stationary-state equation for the ground state and for the states of higher energy.

A comparison with experiments reveals, that $r_{ref}p_{ref}=\hbar$, which I will use in what follows. Thus, with $f_E(r)=r_{ref}^{3/2}\psi_E(r)$ and $g_E(p)=p_{ref}^{3/2}\phi_E(p)$, eqs. (13) are the stationary Schrödinger equations in configuration and momentum configuration spaces.

4. Quantization as selection problem — II. Non-classical solution to the stationary Schrödinger equation

As observed by Schrödinger himself (!), the eigenvalue method used by himself does *not* properly account for the *quantum* nature of quantum systems, because it applies to (and had been developed for) classical systems like strings and pipes. In what follows, I will describe a solution method being free of that deficiency.

4.1. The linear oscillator

The stationary Schrödinger equation for a linear undamped harmonic oscillator reads

$$\hat{H}(x)f_E(\frac{x}{r_{ref}}) = Ef_E(\frac{x}{r_{ref}}); \qquad \hat{H}(x) = \frac{m}{2}\omega^2 x^2 - \frac{\hbar^2}{2m}\frac{\partial^2}{\partial x^2} \tag{14}$$

To see its essentials I introduce dimensionless variables as[5]

$$\xi = \frac{x}{r_{ref}}; \qquad r_{ref} = \sqrt{\frac{r_{ref}p_{ref}}{2m\omega}}; \qquad y_a(\xi) = f_E(\xi); \qquad a = -\frac{E}{r_{ref}p_{ref}\omega} \tag{15}$$

to obtain

$$\frac{d^2 y_a(\xi)}{d\xi^2} - \left(\frac{1}{4}\xi^2 + a\right)y_a(\xi) = 0 \tag{16}$$

This is Weber's equation [48] being one of the equations of the parabolic cylinder [1]. Despite of the reference length, r_{ref}, the stationary states are determined solely through the energy parameter, $-a$. In contrast to the classical oscillator, where $E{\sim}\omega^2$, the quantum oscillator exhibits

---

5 This yields $p_{ref}=2m\omega r_{ref}$; the classical maximum values are interrelated as $p_{max}=m\omega r_{max}$. I deviate from the exposition in [19][10] to make the following clearer.

$E \sim \omega$. Since ω does not occur as a self-standing parameter, the quantization is not affecting it; Schrödinger's 3rd requirement is fulfilled.

4.2. The mathematically distinguished solutions

For and only for the values $a = \pm \frac{1}{2}$ the l.h.s. of eq. (16) factorizes.[6]

$$\left(\frac{d}{d\xi} + \frac{1}{2}\xi\right)\left(\frac{d}{d\xi} - \frac{1}{2}\xi\right) y_{+\frac{1}{2}}(\xi) = 0; \qquad \left(\frac{d}{d\xi} - \frac{1}{2}\xi\right)\left(\frac{d}{d\xi} + \frac{1}{2}\xi\right) y_{-\frac{1}{2}}(\xi) = 0 \qquad (17)$$

Therefore, the values $a = \pm \frac{1}{2}$ are mathematically distinguished against all other a-values. The corresponding solutions, $y_{\pm\frac{1}{2}}(\xi)$, are mathematically equivalent, but physically different. $y_{-\frac{1}{2}}(\xi) = y_{-\frac{1}{2}}(0) \times \exp(-\xi^2/4)$ is a limiting amplitude, while $y_{+\frac{1}{2}}(\xi) = y_{+\frac{1}{2}}(0) \times \exp(+\xi^2/4)$ is not. This distinguishes *physically* the value $a = -\frac{1}{2}$ over the value $a = +\frac{1}{2}$.

If there would be no other distinguished a-values, there would be only *one* state ($a = -\frac{1}{2}$). However, a system having got just *one* state is not able to exchange energy with its environment. In order to find further distinguished a-values, I examine two recurrence relations for the standard solutions of eq. (16) ([1] 19.6.1, 19.6.5).

$$\left(\frac{d}{d\xi} + \frac{1}{2}\xi\right) U(a,\xi) + \left(a + \frac{1}{2}\right) U(a+1,\xi) = 0; \qquad \left(\frac{d}{d\xi} - \frac{1}{2}\xi\right) V(a,\xi) - \left(a - \frac{1}{2}\right) V(a-1,\xi) = 0 \qquad (18)$$

Such recurrence relations can be obtained *without* solving the differential equation, *viz*, from Whittaker's representation of the solutions as contour integrals ([52] 16.61; [1] 19.5). This representation has been developed well before the advance of QM; it is thus independent of the needs of QM.

The recurrence relations

• are not related to the usual, classical solution methods;

• interrelate solution functions with a *finite* difference between their a-values, *viz*, $\Delta a = \pm 1$ (this becomes $\Delta E = \pm \hbar \omega$ later on);

• reflect the *genuine* discrete structure wanted; in particular, this structure has nothing to do with boundary conditions, since *all* solutions exhibit this discrete structure, not only Schrödinger's eigensolutions;

• realize the "conviction", that "the true laws of quantum mechanics would consist not of specific prescriptions for the *single* orbit; but, in these true laws, the elements of the whole manifold of orbits of a system are connected by equations, so that there is apparently a certain interaction between the various orbits." ([43] Second Commun., p. 508)

---

6 These factors are closely related to the variables that factorize the classical Hamiltonian.

Moreover, the recurrence relations divide the set of a-values as follows.

Set (1) $a =..., -5/2, -3/2, -1/2$; the 2nd relation (18) breaks at $a=-\frac{1}{2}$ being one of the two mathematically distinguished values found above;

Set (2) $a =..., 5/2, 3/2, 1/2$; the 1st relation (18) breaks at $a=+\frac{1}{2}$ being the other mathematically distinguished value found above;

Set (3) $a = \{..., -2+\varsigma, -1+\varsigma, \varsigma, \varsigma+1, \varsigma+2... | -\frac{1}{2} < \varsigma < +\frac{1}{2}\}$; there is no break in the recurrence relations (18) for this set.

The smallest interval representing *all* solutions is the closed interval $a=[-\frac{1}{2},+\frac{1}{2}]$, all other solutions being related to it through the recursion formulae. The values $a=-\frac{1}{2}$ (set (1)) and $a=+\frac{1}{2}$ (set (2)) are mathematically distinguished, again; this time as the boundary points of that interval. All inner interval points, $-\frac{1}{2}<a<+\frac{1}{2}$ (set (3)), are mathematically equivalent among each another and, consequently, not distinguished. The physically relevant set of a-values is a mathematically distinguished set.

4.3. The physically distinguished solutions

The mathematically distinguished set (1) contains the physically relevant value $a=-\frac{1}{2}$, while the mathematically distinguished set (2) contains the unphysical value $a=+\frac{1}{2}$. The recurrence relations (18) show, that all functions $U(a,\xi)$ with a-values from set (1) are limiting amplitudes, while all functions $V(a,\xi)$ with a-values from set (2) are not. For the a-values of set (3), neither $U(a,\xi)$, nor $V(a,\xi)$ is a limiting amplitude.

Moreover, set (1) exhibits a *finite minimum* of total energy, $E=\frac{1}{2}r_{ref}p_{ref}\omega$, while sets (2) and (3) do not. A system of sets (2) or (3) can deliver an unlimited amount of energy to its environment, it is a *perpetuum mobile* of 1st kind. This makes set (1) to be physically distinguished against sets (2) and (3).

Hence, starting from the ground state, $y_{-\frac{1}{2}}(\xi)=y_{-\frac{1}{2}}(0)\times\exp(-\xi^2/4)$, and using the recursion formula (18) for $U(a,\xi)$, the physically relevant solutions are obtained as

$$y_{-n-\frac{1}{2}}(\xi) = e^{-\frac{1}{4}\xi^2} He_n(\xi); \qquad n = 0,1,2,... \tag{19}$$

where

$$He_n(\xi) = (-1)^n e^{\frac{1}{2}\xi^2} \frac{d^n}{d\xi^n} e^{-\frac{1}{2}\xi^2} \tag{20}$$

is the nth Hermite polynomial ([1] 19.13.1). Schrödinger's boundary condition (the wave function should vanish at infinity) is fulfilled automatically.

Since ([1] 22.2)

$$\int\limits_{-\infty}^{+\infty} e^{-\frac{1}{2}\xi^2} He_n(\xi)^2 d\xi = \sqrt{2\pi}n! \tag{21}$$

the normalized solutions [see eq. (8)] read

$$y_{-n-\frac{1}{2}}(\xi) = \frac{1}{\sqrt{\sqrt{2\pi}n!}} e^{-\frac{1}{4}\xi^2} He_n(\xi); \qquad n = 0,1,2,\dots \tag{22}$$

4.4. The non-classical potential energy and the tunnel effect

The observation of quantum particles crossing spatial domains, where $V>E$, has led to the notion 'tunnel effect' [34][38]. Being a nice illustration, this wording masks the fact, that the actual contribution of a configuration, $\mathbf{r}$, to the total energy is not $V(\mathbf{r})$, but $V_E^{ncl}(\mathbf{r})<E$, see eq. (6).

In terms of the dimensionless variables (15) the dimensionless non-classical potential energy of the oscillator above equals

$$v_n^{ncl}(\xi) = y^2_{-n-\frac{1}{2}}(\xi)\frac{1}{2}\xi^2 = \frac{1}{2\sqrt{2\pi}n}\xi^2 e^{-\frac{1}{2}\xi^2} He_n(\xi)^2 \tag{23}$$

Using the recurrence formula $\xi He_n(\xi) = He_{n+1}(\xi) + n He_{n-1}(\xi)$ ([1] 22.7.14) and the inequality $|He_n(\xi)| < \exp(\xi^2/4)\sqrt{(n!)}k$, $k \approx 1.086435$ (ibid., 22.14.17), one can prove, that

$$v_n^{ncl}(\xi) = y_{-n-\frac{1}{2}}(\xi)^2 \frac{1}{2}\xi^2 < n + \frac{1}{2}; \qquad -\infty < \xi < +\infty; \qquad n = 0,1,2,\dots \tag{24}$$

Hence, the inequalities (6) are fulfilled.

The occurrence of the 'smaller than' sign means, that - in contrast to the classical oscillator - there are no stationary states with, (i), vanishing potential energy (in particular, the ground state is not a state at rest) and, (ii) vanishing kinetic energy (there are no turning points).

The picture of the tunnel is partially correct, in that the classical turning points are points of inflection such, that beyond them, in the forbidden domains of Newtonian CM, the wave function decreases exponentially.

Notice that these results follow solely from the most general principles of state description according to Leibniz [35], Euler, Helmholtz and Schrödinger, without solving any stationary-state equation or equation of motion and without assuming particular boundary conditions.

5. The time dependent case

While Heisenberg [30] and Schrödinger [43] started from a time-dependent equation[7], I have worked so far with the set of all possible (momentum) configurations of systems in their stationary states, where time plays no role. In order to incorporate time, I will proceed as Newton and Euler did in the classical case and will consider first the time-dependence of the stationary states, then, the change of these states, and, finally, I will arrive at the time-dependent Schrödinger equation as the equation of motion.

5.1. The time dependence of the stationary states

According to their definitions (6), the stationary weight functions, $F_E(\mathbf{r})$ and $G_E(\mathbf{p})$, are time independent. Hence, if there is a time dependence of the stationary weight amplitudes, and correspondingly of the wave functions, it is of the form

$$\psi_E(\mathbf{r},t) = e^{i\phi_E(t)}\psi_E(\mathbf{r}); \qquad \varphi_E(\mathbf{p},t) = e^{i\phi_E(t)}\varphi_E(\mathbf{p}) \tag{25}$$

The phase, $\varphi_E(t)$, is the same for both functions, since the Fourier transform (9) is time-independent.

For a free particle,

$$\psi_E(\mathbf{r},t) \sim \exp\{i\mathbf{k}_E \cdot \mathbf{r} - i\omega_E t\}; \qquad E = \frac{\hbar^2 k_E^2}{2m} \tag{26}$$

The group velocity equals the time-independent particle velocity.

$$\mathbf{v}_g = \frac{d\omega_E}{d\mathbf{k}_E} = \frac{\hbar \mathbf{k}_E}{m} \quad \Rightarrow \quad \omega_E = \frac{\hbar k_E^2}{2m} = \frac{E}{\hbar} \tag{27}$$

Therefore,

$$\psi_E(\mathbf{r},t) = e^{-i\frac{E}{\hbar}t}\psi_E(\mathbf{r}); \qquad \phi_E(\mathbf{p},t) = e^{-i\frac{E}{\hbar}t}\phi_E(\mathbf{p}) \tag{28}$$

For later use I remark that this can be written as

---

7 In fn. 2, p. 489, of the 2nd Commun., Schrödinger has distanced himself from the time-independent approach of the 1st Commun.

$$\psi_E(\mathbf{r},t) = e^{-\frac{i}{\hbar}\hat{H}(\mathbf{r})t}\psi_E(\mathbf{r}); \qquad \phi_E(\mathbf{p},t) = e^{-\frac{i}{\hbar}\hat{H}(\mathbf{p})t}\phi_E(\mathbf{p}) \tag{29}$$

5.2. The equation-of-state-change

The analog to the Hamiltonian as classical stationary-state function is the function

$$Z^{ncl} = \frac{\iiint\limits_{C^{all}} \overline{f}(\mathbf{r},t)\hat{H}(\mathbf{r},t)f(\mathbf{r},t)d^3r}{\iiint\limits_{C^{all}} \overline{f}(\mathbf{r},t)f(\mathbf{r},t)d^3r} = \frac{\iiint\limits_{P^{all}} \overline{g}(\mathbf{p},t)\hat{H}(\mathbf{p},t)g(\mathbf{p},t)d^3p}{\iiint\limits_{P^{all}} \overline{g}(\mathbf{p},t)g(\mathbf{p},t)d^3p} \tag{30}$$

The analog to the classical principles of change of state, CS1...CS5, reads as follows. Up to first order in dt,

QS1) the change of stationary-state quantities (dZ) depends solely on the external causes (Z_{ext}), but not on state-of-motion quantities (f, g);

QS2) the changes of the stationary-state quantities (dZ) are independent of the stationary-state quantities (Z) themselves;

QS3) the changes of state-of-motion quantities (df, dg) depend directly solely on stationary-state quantities (Z);

QS4) the changes of stationary-state (dZ) and of state-of-motion quantities (df, dg) are independent each of another;

QS5) as soon as the external causes (Z_{ext}) vanish, the system remains in the (not necessarily stationary[8]) state assumed in this moment.

Hence, writing

$$\iiint \overline{f}(\mathbf{r},t)\hat{H}(\mathbf{r},t)f(\mathbf{r},t)d^3r = \left\langle f(\mathbf{r},t)\left|\hat{H}(\mathbf{r},t)\right|f(\mathbf{r},t)\right\rangle \tag{31}$$

the equation of state change becomes

$$
\begin{aligned}
dZ^{ncl} &= \frac{\left\langle df(\mathbf{r},t)\left|\hat{H}(\mathbf{r},t)\right|f(\mathbf{r},t)\right\rangle + \left\langle f(\mathbf{r},t)\left|d\hat{H}(\mathbf{r},t)\right|f(\mathbf{r},t)\right\rangle + \left\langle f(\mathbf{r},t)\left|\hat{H}(\mathbf{r},t)\right|df(\mathbf{r},t)\right\rangle}{\left\langle f(\mathbf{r},t)\left|f(\mathbf{r},t)\right\rangle\right.} \\
&\quad - \frac{\left\langle df(\mathbf{r},t)\left|f(\mathbf{r},t)\right\rangle + \left\langle f(\mathbf{r},t)\right|df(\mathbf{r},t)\right\rangle}{\left\langle f(\mathbf{r},t)\left|f(\mathbf{r},t)\right\rangle\right.} \frac{\left\langle f(\mathbf{r},t)\left|\hat{H}(\mathbf{r},t)\right|f(\mathbf{r},t)\right\rangle}{\left\langle f(\mathbf{r},t)\left|f(\mathbf{r},t)\right\rangle\right.} \\
&\stackrel{!}{=} \frac{\left\langle f(\mathbf{r},t)\left|d\hat{H}(\mathbf{r},t)\right|f(\mathbf{r},t)\right\rangle}{\left\langle f(\mathbf{r},t)\left|f(\mathbf{r},t)\right\rangle\right.}
\end{aligned} \tag{32}
$$

---

8 The modification against CS5 is a consequence of the discreteness of the energetic spectrum.

5.3. Derivation of the time-dependent Schrödinger equation

The requirement in eq. (32) implies two conditions.

$$\left\langle df(\mathbf{r},t)\middle|\hat{H}(\mathbf{r},t)\middle|f(\mathbf{r},t)\right\rangle + \left\langle f(\mathbf{r},t)\middle|\hat{H}(\mathbf{r},t)\middle|df(\mathbf{r},t)\right\rangle = 0$$
$$\left\langle df(\mathbf{r},t)\middle|f(\mathbf{r},t)\right\rangle + \left\langle f(\mathbf{r},t)\middle|df(\mathbf{r},t)\right\rangle = d\left\langle f(\mathbf{r},t)\middle|f(\mathbf{r},t)\right\rangle = 0 \tag{33}$$

The second condition means that there is a unitary time development operator,

$$\hat{U}(\mathbf{r},t_2,t_1)f(\mathbf{r},t_1) = f(\mathbf{r},t_2); \qquad \hat{U}(\mathbf{r},t_2,t_1)^\dagger = \hat{U}(\mathbf{r},t_2,t_1)^{-1} \tag{34}$$

such, that

$$\left\langle f(\mathbf{r},t_2)\middle|f(\mathbf{r},t_2)\right\rangle = \left\langle \hat{U}(\mathbf{r},t_2,t_1)f(\mathbf{r},t_1)\middle|\hat{U}(\mathbf{r},t_2,t_1)f(\mathbf{r},t_1)\right\rangle = \left\langle f(\mathbf{r},t_1)\middle|f(\mathbf{r},t_1)\right\rangle \tag{35}$$

Now I insert eq. (34) into the first requirement (33).

$$\left\langle d\hat{U}(\mathbf{r},t,0)f(\mathbf{r},0)\middle|\hat{H}(\mathbf{r},t)\middle|\hat{U}(\mathbf{r},t,0)f(\mathbf{r},0)\right\rangle + \left\langle \hat{U}(\mathbf{r},t,0)f(\mathbf{r},0)\middle|\hat{H}(\mathbf{r},t)\middle|d\hat{U}(\mathbf{r},t,0)f(\mathbf{r},0)\right\rangle = 0 \tag{36}$$

The unitary solution to this equation reads $d\hat{U}(r, t, 0) = iu(\hat{H}(r, t))dt$, where $u(\hat{H})$ is a real-valued rational function of the self-adjoint Hamiltonian, H. Compatibility with the stationary case (29) yields $u(\hat{H}) = -\dfrac{1}{\hbar}\hat{H}$. Hence,

$$\hat{U}(\mathbf{r},t,0) = \hat{P}\left(\exp\left\{ \frac{-i}{\hbar}\int_0^t \hat{H}(\mathbf{r},t')dt' \right\} \right) \tag{37}$$

where $\hat{P}$ denotes Dyson's time-ordering operator [4]. The time-dependent Schrödinger equation for $f(\mathbf{r},t)$ follows immediately.

The momentum representation can be derived quite analogously.

Both representations of the time-dependent Schrödinger form two equivalent equations of motion. As in the classical case, the equation of motion is a dynamic equation for non-stationary-state entities.

6. Summary and conclusions

I have presented a relatively novel approach to quantization, *viz*, quantization as selection rather than eigenvalue problem. It starts from Euler's rather than Newton's axiomatic and exploits Helmholtz's [32][33] treatment of the energy conservation law. It fulfills all four of Schrödinger's methodical requirements quoted in the Introduction.

It is often assumed, that the difference between classical and quantum systems is caused by the existence of the quantum of action. I have shown that this assumption is not necessary. It is sufficient to make different assumptions about the set of (momentum) configurations a mechanical system can assume in its stationary states.

The "logical opposition" between CM and QM observed by Schrödinger [44] is actually a dialectic relationship, which resembles that between the Finite and the Infinite. Each determination draws a Limit, where each Limit involves the existence of something beyond it (cf [29] Logik I p. 145). The notion of the Finite does not exist without the notion of the Infinite (ibid. pp. 139ff.). The Infinite is the Other of the Finite - in turn, the Finite is the Other of the Infinite. Now, the Finite and the Infinite are not simply opposites; a border between them would contradict the very meaning of infinity. The True Infinite includes the Finite, it is the unity of the Finite and the Infinite (cf ibid. p. 158).

The solution of the stationary Schrödinger equation without using boundary conditions shows, that it actually does "carry the quantum conditions in itself" (Schrödinger's 1st requirement, see Introduction). Hence, it has got "maximum strength" in the sense of Einstein [8].

CM contains the necessary means for going beyond its own frame. This way, the relationship between CM and non-CM becomes well defined, and the physical content of non-CM is formulated on equal footing with the mathematical method (and vice versa). An example for this is the reformulation of Einstein's criterion (the number of stationary states) in terms of recurrence relations.

Ad-hoc assumptions, which may be suggested by experimental results, but are not supported by the axiomatic of CM, can be avoided. The wave and particle aspects can be obtained from the time-dependent Schrödinger equation and its solutions [10]. The classical path in phase space is replaced with the wave functions in space and momentum representations. The wave functions take also the role of the initial conditions, which "are not free, but also have to obey certain laws" [7].

The dynamics in space and in momentum space are treated in parallel. As a consequence, the Schrödinger equation in momentum representation is obtained at once with the one in position representation. This, too, enables one to keep maximum contact to CM and to explain, why QM is a non-classical mechanics of conservative systems, where the classical potential and kinetic energy functions and, consequently, the classical Lagrange and Hamilton functions still apply. This includes a natural explanation of "the peculiar significance of the energy in quantum mechanics" [51].

Modern representations of CM favor equations of motion as the foundation (the variational principles belong to this class). The state variables are position and velocity (Lagrange, Laplace), or position and momentum (Hamilton). Hence, there are 6 state variables for a single body. In contrast, there are only 3 quantum numbers for a spinless particle. And there are only 3 stationary-state variables for a single body within Newton's (the 3 components of the momentum vector) and Euler's (the 3 components of the velocity vector) representations of CM, respectively.[9] This is another indication for the fact, that the latter are more suitable for the transition from CM to QM than Lagrange's and Hamilton's representations.

For the quantization of fields, finally, this approach yields an explanation for the fact, that, within the method of normal-mode expansion, only the temporal, but not the spatial part of the field variables is concerned (cf [42]). Indeed, only those variables are subject to the quantization procedure, the possible values of which are restricted by the energy law. The spatial extension of the normal modes is fixed by the boundary conditions and thus not subject to quantization. The classical field energy (density) is determined by the normal-mode amplitudes and thus limits these. As a consequence, the time-dependent coefficients in the normal-mode expansion are quantized. When formulating this expansion such, that these expansion coefficients get the dimension of length, their quantization can be performed in complete analogy to that of the harmonic oscillator, without invoking additional assumptions or new constants [10]. Moreover, one could try to quantize a field in the space spanned by independent dynamical field variables. This could separate the quantization problem from the spatial and temporal field distributions and, thus, simplify the realization of Einstein's imagination of a "spatially granular" [5] structure of the electromagnetic field.

Acknowledgements

I feel highly indebted to Dr. D. Suisky with whom the basic ideas of 'quantization as selection problem' have been elaborated [46][47][18][19]. Over the years I have benefited from numerous discussions with Dr. M. Altaisky, Prof. Y. Dabaghian, Dr. M. Daumer, Dr. K. Ellmer, Dr. D. B. Fairlie, Dr. A. Förster, Prof. L. Fritsche, Prof. W. Greiner, Dr. H. Hecht, H. Hille, Prof. J. Keller, Prof. J. R. Klauder, Prof. H. Kröger, Dr. Th. Krüger, Prof. H. Lübbig, Prof. G. Mann, Prof. Matone, Prof. S. N. Mayburow, Prof. P. Mittelstaedt, Dr. R. Müller, Prof. J. G. Muroz, Prof. G. Nimtz, Prof. H. Paul, Prof. Th. Pöschel, Prof. J. Rosen, A. Rothenberg, Prof. W. P. Schleich, Prof. J. Schröter, Prof. J. Schnakenberg, Prof. J. Schröter, Dr. W. Smilga, Dr. E. V. Stefanovich, Dr. L. Teufel, Dr. R. Tomulka, Prof. H. Tributsch, Dr. M. Vogt, Prof. R. F. Werner and many more. I am indebted to Profs. M. Müller-Preußker and W. Nolting for their continuous interest and support. Early stages of this work were supported by the Deutsche Akademie der Naturforscher Leopoldina [18], Prof. Th. Elsässer and Prof. E. Siegmund.

Author details

Peter Enders*

Address all correspondence to: enders@dekasges.de

University of Applied Sciences, Wildau, Königs Wusterhausen, Germany

---

9 Some implications of this similarity between Newton's and Euler's notions of state on the classical and Schrödinger's and Pauli's [39][40] on the quantum sides have been investigated in [9][11][12][15][16].

References

[1] Abramowitz, M, & Stegun, I. A. Eds.), *Handbook of Mathematical Functions*, Washington: NBS 1964; abbreviated reprint (selection by M. Danos & J. Rafelski): *Pocketbook of Mathematical Functions*, Thun Frankfurt/Main: Deutsch (1984).

[2] Birkhoff, G, & Von Neumann, J. *The logic of quantum mechanics*, Ann. of Math. [2] (1936)., 37(1936), 823-843.

[3] Bohr, N. (1913). On the Constitution of Atoms and Molecules", Phil. Mag., 26, 1-13.

[4] Dyson, F. J. (1949). The radiation theories of Tomonaga, Schwinger, and Feynman", Phys. Rev., 75, 486-502.

[5] Einstein, A. (1905). Über einen die Erzeugung und Verwandlung des Lichtes betreffenden heuristischen Gesichtspunkt", Ann. Phys., 17, 132-148.

[6] Einstein, A. (1907). Die Plancksche Theorie der Strahlung und die Theorie der spezifischen Wärme", Ann. Phys. 22, , 180ff.

[7] Einstein, A. (1923). Bietet die Feldtheorie Möglichkeiten für die Lösung des Quantenproblems?", Sitzungsber. Preuss. Ak. Wiss. phys. math. Kl., 13. Dez,, XXXIII, , 359ff.

[8] Einstein, A. (1977). *Grundzüge der Relativitätstheorie* (Akademie-Verlag, Berlin), , 132.

[9] Enders, P. (2004). *Equality and Identity and (In)distinguishability in Classical and Quantum Mechanics from the Point of View of Newton's Notion of State*, 6th Int. Symp. Frontiers of Fundamental and Computational Physics, Udine; in: Sidharth, Honsell & De Angelis (Hrsg.), *Frontiers of Fundamental Physics*, 2006, , 239-245.

[10] Enders, P. (2006). *Von der klassischen Physik zur Quantenphysik. Eine historisch-kritische deduktive Ableitung mit Anwendungsbeispielen aus der Festkörperphysik*, Berlin Heidelberg: Springer

[11] Enders, P. *Is Classical Statistical Mechanics Self-Consistent?* (A paper of honour of C. F. von Weizsäcker, 1912-2007), Progr. Phys. (2007). http://www.allbusiness.com/science-technology/physics/5518225-1.html, 3(2007), 85-87.

[12] Enders, P. *Equality and Identity and (In)distinguishability in Classical and Quantum Mechanics from the Point of View of Newton's Notion of State*, Icfai Univ. J. Phys. I ((2008). http://www.iupindia.org/108/IJP_Classical_and_Quantum_Mechanics_71.html

[13] Enders, P. *Towards the Unity of Classical Physics*, Apeiron 16 ((2009). http://redshift.vif.com/JournalFiles/V16N1END.pdf

[14] Enders, P. *Huygens principle as universal model of propagation*, Latin Am. J. Phys. Educ. (2009). http://dialnet.unirioja.es/servlet/articulo?codigo=3688899, 3(2009), 19-32.

[15] Enders, P. *Gibbs' Paradox in the Light of Newton's Notion of State*, Entropy (2009). http://www.mdpi.com/1099-4300/11/3/454, 11(2009), 454-456.

[16] Enders, P. *State, Statistics and Quantization in Einstein's 1907 Paper,`Planck's Theory of Radiation and the Theory of Specific Heat of Solids'*, Icfai Univ. J. Phys. II ((2009). http:// www.iupindia.org/709/IJP_Einsteins_1907_Paper_176.html

[17] Enders, P. (2010). Precursors of force fields in Newton's 'Principia'", Apeiron 17, 22-27; http://redshift.vif.com/JournalFiles/V17N1END.PDF

[18] Enders, P, & Suisky, D. (2004). Über das Auswahlproblem in der klassischen Mechanik und in der Quantenmechanik", Nova Acta Leopoldina, Suppl. , 18, 13-17.

[19] Enders, P, & Suisky, D. (2005). Quantization as selection problem", Int. J. Theor. Phys. , 44, 161-194.

[20] Euler, L. II-1, "Mechanica sive motus scientia analytice exposita", in: Leonardi Euleri *Opera Omnia sub auspiciis Societatis Scientarium Naturalium Helveticae* (Zürich Basel, 1911 - 1986), ser. II, vol. 1

[21] Euler, L, & Découverte, I. I-5a. d'une nouveau principe de mécanique", in: *Opera Omnia*, ser. II, , 5

[22] Euler, L. II-5b, "Harmonie entre les principes généraux de repos et de mouvement de M. de Maupertuis", in: *Opera Omnia*, ser. II, , 5

[23] Euler, L. II-5c, "Recherches sur l'origine des forces", Mém. ac. sci. Berlin 6 (1750) 1752, 419-447; in: *Opera Omnia*, ser. II, , 5

[24] Euler, L. III-1, "Anleitung zur Naturlehre", in: *Opera Omnia*, ser. III, , 1

[25] Euler, L. III-11, "Lettres à une princesse d'Allemagne sur divers sujets de Physique et de Philosophie", in: *Opera Omnia*, ser. III, vols. 11 and 12

[26] Faraggi, A. E, & Matone, M. (1998). The Equivalence Postulate of Quantum Mechanics", arXiv:hep-th/9809127

[27] Feynman, R. P. (1949). Space-Time Approach to Quantum Electrodynamics", Phys. Rev., 76, , 769ff.

[28] FrancisCh. (2010). *Quantum Logic*, http://rqgravity.net/FoundationsOfQuantumTheory#QuantumLogic

[29] (G. F. W. Hegel, Werke in 20 Bänden, Frankfurt/M.: Suhrkamp 1969-1971 (stw 601-620).

[30] Heisenberg, W. (1925). Über quantenmechanische Umdeutung kinematischer und mechanischer Beziehungen", Z. Phys. XXXIII, , 879ff.

[31] Heisenberg, W. (1977). Die Geschichte der Quantentheorie", in: *Physik und Philosophie* (Ullstein, Frankfurt etc), , 15-27.

[32] Helmholtz, H. (1847). *Über die Erhaltung der Kraft* (Reimer, Berlin)

[33] Helmholtz, H. von (1911). *Vorlesungen über die Dynamik discreter Massenpunkte* (Barth, Leipzig, 2nd ed.)

[34] Hund, F. *Zur Deutung der Molekelspektren. I*, Z. Phys. (1927). *III. Bemerkungen über das Schwingungs- und Rotationsspektrum bei Molekeln mit mehr als zwei Kernen*, 43 (1927) 805-826, 40(1927), 742-764.

[35] Leibniz, G. W. *Specimen dynamicum pro admirandis naturae legibus circa corporum vires et mutuas actiones detegendis et ad suas causas revocandis*, Acta erudit. Lipsiens. April ; Engl.: *Essay in Dynamics showing the wonderful laws of nature concerning bodily forces and their interactions, and tracing them to their causes*, http://www.earlymodern-texts.com/pdf/leibessa.pdftranslated, edited and commented by J. Bennett, June (2006).

[36] Messiah, A. (1999). *Quantum Mechanics* (Dover, New York), § I.15

[37] Newton, I. (1999). *The Principia. Mathematical Principles of Natural Philosophy* (A New Translation by I. Bernhard Cohen and Anne Whitman assisted by Julia Buden, Preceded by *A Guide to Newton's Principia* by I. Bernhard Cohen), Berkeley etc.: Univ. Calif. Press

[38] Nordheim, L. W. *Zur Theorie der thermischen Emission und der Reflexion von Elektronen an Metallen*, Z. Phys. (1928) , 46(1928), 833-855.

[39] Pauli, W. (1926). Quantentheorie", in: H. Geiger and K. Scheel (Eds.): *Handbuch der Physik* (Springer, Berlin), , 23, 1-278.

[40] Pauli, W. (1973). *Wave Mechanics* (Pauli Lectures of Physics, Ed. Ch. P. Enz (MIT Press, Cambridge, Mass., 1973), § 35, 5

[41] Planck, M. (1900). Zur Theorie des Gesetzes der Energieverteilung im Normalspektrum", Verh. Dtsch. Phys. Ges., 2, , 237ff.

[42] Schleich, W. P. (2001). *Quantum Optics in Phase Space* (Berlin etc., Wiley-VCH), , 282.

[43] Schrödinger, E. (1926). *Quantisierung als Eigenwertproblem. Erste Mitteilung*, Ann. Phys. [4] *Zweite Mitteilung*, 489-527; *Dritte Mitteilung: Störungstheorie, mit Anwendung auf den Starkeffekt der Balmerlinien*, 80 (1926) 437-490; *Vierte Mitteilung*, 81 (1926) 109-139; reprints in: *Abhandlungen zur Wellenmechanik* (Barth, Leipzig), 79(1926), 361-376.

[44] Schrödinger, E. (1933). The fundamental idea of wave mechanics" (Nobel award lecture, 1933)

[45] Suisky, D. (2009). *Euler as Physicist*, Berlin Heidelberg: Springer

[46] Suisky, D, & Enders, P. (2001). Leibniz's Foundation of Mechanics and the Development of 18th Century Mechanics initiated by Euler", in: H. Poser (Ed.), *Nihil sine ratione*, Proc. VII Intern. Leibniz Congress, Berlin, http://www.leibniz-kongress.tu-

berlin.de/webprogramm.html;http://www.information-philosophie.de/philosophie/leibniz2001.html, 1247.

[47] Suisky, D, & Enders, P. (2003). On the derivation and solution of the Schrödinger equation. Quantization as selection problem", Proc. 5th Int. Symp. Frontiers of Fundamental Physics, Hyderabad (India), Jan. , 8-11.

[48] Weber, H. *Über die Integration der partiellen Differentialgleichung...*, Math. Ann. I ((1869).

[49] Weisstein, E. W. (2012). Fourier Transform", From MathWorld-A Wolfram Web Resource; http://mathworld.wolfram.com/FourierTransform.htmlMarch 13, 2012)

[50] Weizsäcker, C. F. v. (2002). *Aufbau der Physik* (dtv, München, 4th ed.), 235.

[51] Weyl, H. (1950). *The Theory of Groups in Quantum Mechanics* (Dover, New York), § II.8

[52] Whittaker, E. T, & Watson, G. N. *A Course of Modern Analysis,* Cambridge: Cambridge Univ. Press '1927, new ed. (1996). (Cambr. Math. Libr. Ser.)

Chapter 3

The Husimi Distribution: Development and Applications

Sergio Curilef and Flavia Pennini

Additional information is available at the end of the chapter

1. Introduction

The Husimi distribution, introduced by Kôdi Husimi in 1940 [1], is a quasi-probability distribution commonly used to study the correspondence between quantum and classical dynamics [2]. Also, it is employed to describe systems in different areas of physics such as Quantum Mechanics, Quantum Optics, Information Theory [3–8]. Additionally, in nanotechnology it is possible to obtain a clear description of localization –which corresponds to classicality– and is crucial to determine correctly the size of systems when the particle dynamics takes into account mobility boundaries [9]. Among its properties, it is always positive definite and unique, conversely it cannot be considered as a true probability distribution over the quantum-mechanical phase space, reason why it is often considered as a quasi probability distribution. Although it possesses no correct marginal properties, its usefulness is to allow the assessment of the expectation values in quantum mechanics in a way similar to the classical case [10]. The semiclassical Husimi probability distribution refers to a special type of probability, this is for simultaneous but approximate location of position and momentum in phase space.

The Husimi distribution may be obtained in several ways; the strategy that we choose here is to derive it as the expectation value of the density operator in a basis of coherent states [11]. Therefore, the line of working in this chapter is illustrated in the following sequence:

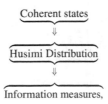

Coherent states
⇓
Husimi Distribution
⇓
Information measures,

where the transcendence of defining correctly a set of coherent states and the Husimi distribution is evident, being the calculation of measures as Wehrl entropy and/or Fisher information a consequence of this procedure.

Coherent states provide a close connection between classical and quantum formulations of a given system. They were introduced early by Erwin Schrödinger in 1926 [12], but the name *coherent state* appeared for first time in Glauber's papers [13, 14]– see a detailed study about this in Ref. [15]. It is known that is difficult to construct coherent states for arbitrary quantum mechanical systems. Klauder shows an elegant method for construct it in Ref. [16]. Furthermore, in Ref. [11] Gazeau and Klauder consider essential, among other things, to discuss what an appropriate formulation of coherent states needs [11]. For instance, they suggest a suitable set of requirements. Then, the main interest in this chapter is to discuss, starting from a well defined set of coherent states, some interesting problems related to the Husimi distribution applied to important systems in physics, such as, the harmonics oscillator [5], the Landau diamagnetism model [17, 18] and, the rigid rotator [6, 18]. Also, we will discuss some properties related to systems with continuous spectrum [19]. In each case, the Wehrl entropy is calculated as a possible application.

This chapter is organized as follows. In section 2 we start presenting the background material and methodology that will be employed in the following chapters. In section 3 we revise the Husimi distribution and the Wehrl entropy for the problem of a particle in a magnetic field. In section 4 we discuss phase space delocalization for the rigid rotator within a semiclassical context by recourse to the Husimi distributions of both the linear and the $3D-$anisotropic instances. In section 5 we propose a procedure to generalize the Husimi distribution to systems with continuous spectrum. We start examining a pioneering work, by Gazeau and Klauder, where the concept of coherent states for systems with discrete spectrum was extended to systems with continuous one. Finally, some concluding remarks and open problems are commented in section 6 .

2. Background material and methodology

In this section we center our attention in 3 topics that we consider relevant to understand the problems that will be discussed in the following sections. These are *i)* the Husimi distribution and the most direct application, i.e., Wehrl entropy, *ii)* a special basis to formulate a suitable set of coherent states and *iii)* a generalization of this concepts to systems with continuous spectrum.

2.1. Husimi distribution and Wehrl entropy

The standard statistical mechanics starts conventionally using the Gibbs's canonical distribution, whose thermal density matrix is represented by

$$\hat{\rho} = Z^{-1} e^{-\beta \hat{H}}, \tag{1}$$

where $Z = \text{Tr}(e^{-\beta \hat{H}})$ is the partition function, $\hat{H}$ is the Hamiltonian of the system, $\beta = 1/k_B T$ the inverse temperature T, and k_B the Boltzmann constant [20].

The Husimi distribution is obtained as the expectation value of the density operator in a basis of coherent states as follows [1]

$$\mu(z) = \langle z | \hat{\rho} | z \rangle, \tag{2}$$

where $\{|z\rangle\}$ denotes the set of coherent states, which are the eigenstates of the annihilation operator $\hat{a}$, i.e., $\hat{a}|z\rangle = z|z\rangle$ defined for all $z \in \mathbb{C}$ [11]. This distribution is normalized to unity according to

$$\frac{1}{\pi} \int d^2z \mu(z) = 1, \tag{3}$$

where the integration is carried out over the complex z plane and the differential is a real element of area proportional to phase space element given by $d^2z = dxdp/2\hbar$.

For an arbitrary Hamiltonian $\hat{H}$, with the discrete spectra $\{E_n\}$, being n a positive integer, the Husimi distribution takes the form

$$\mu(z) = \frac{1}{Z} \sum_n e^{-\beta E_n} |\langle z|n\rangle|^2, \tag{4}$$

where $\{|n\rangle|\}$ is the set of energy eigenstates with eigenvalues E_n [4, 5].

The Wehrl entropy is a direct application that we introduce here, which is a useful measure of localization in phase-space [21, 22], whose pertinent definition reads

$$W = -\frac{1}{\pi} \int d^2z \mu(z) \ln\mu(z), \tag{5}$$

The uncertainty principle manifests itself through the inequality $W \geq 1$ which was first conjectured by Wehrl [21] and later proved by Lieb (see, for instance Ref. [4]).

In the special case of the Harmonic Oscillator –whose Hamiltonian is $\hat{H} = \hbar\omega[\hat{a}^\dagger\hat{a} + 1/2]$– its set of Glauber's coherent states is defined in the form [14]

$$|z\rangle = e^{-|z|^2/2} \sum_{n=0}^{\infty} \frac{z^n}{\sqrt{n!}} |n\rangle, \tag{6}$$

where $\{|n\rangle\}$ are a complete orthonormal set of phonon-eigenstates, that is,

$$\langle n|n'\rangle = \delta_{n,n'} \tag{7}$$

where $\delta_{n,n'}$ is the Kronecker delta function, and the energy-spectrum is given by $E_n = \hbar\omega(n + 1/2)$, with $n = 0, 1, \ldots$ By definition, Hermitian operator $\hat{H}$ is an observable if this orthonormal system of vectors forms a basis in the state space. This can be expressed by the closure relation

$$\sum_{n=0}^{\infty} |n'\rangle\langle n| = \hat{1}, \tag{8}$$

where $\hat{1}$ stands for the identity operator in the space formed by eigenvectors.

In this situation one conveniently resorts to

$$\mu_{HO}(z) = (\hat{1} - e^{-\beta\hbar\omega})\,e^{-(1-e^{-\beta\hbar\omega})|z|^2}, \tag{9}$$

$$W_{HO} = 1 - \ln(1 - e^{-\beta\hbar\omega}). \tag{10}$$

which respectively are the useful analytical expressions for Husimi distribution and Wehrl entropy [4].

2.2. Gazeau and Klauder's coherent states

Now, we go back to the set of coherent states defined in Eq. (6). Certainly, it is known that coherent states can be constructed in several ways by recourse to different techniques being its formulation of a not unique character. Nevertheless, contrary to this idea and in order to get a unifying perspective, Gazeau and Klauder have suggested that a suitable formalism for coherent states should satisfy at least the following requirements [11]:

1. *Continuity of labeling* refers to the map from the label space L into Hilbert space. This condition means that the expression $\| \,|z'\rangle - |z\rangle \, \| \to 0$ whenever $z' \to z$.

2. *Resolution of Unity*: a positive measure $\tau(z)$ on L exists such that the unity operator admits the representation

$$\int_L |z\rangle\langle z|\,d\tau(z) = 1, \tag{11}$$

   where $|z\rangle\langle z|$ denotes a projector, which takes a state vector into a multiple of the vector $|z\rangle$.

3. *Temporal Stability*: the evolution of any coherent state $|z\rangle$ always remains a coherent state, which leads to a relation of the form

$$|z(t)\rangle = e^{-i\hat{H}t/\hbar}|z\rangle, \tag{12}$$

   where $z(0) = z$, for all $z \in L$ and t.

4. *Action Identity*: this property requires that

$$\langle z|\hat{H}|z\rangle = \hbar\omega|z|^2. \tag{13}$$

At this point, we remark that requirements (3) and (4) are directly satisfied when the spectrum of the Hamiltonian $\hat{H}$ of the system, has the form $E_n \sim n\hbar\omega$, where n is the quantum number and ω is the frequency of the oscillator [11]. In addition, there are some shortcomings about these requirements; for instance, Gazeau and Klauder states cannot be used for degenerate systems. Furthermore, it is questionable that action identity leads to the classical action-angle variable interpretation [23].

2.3. Continuous spectrum

Gazeau and Klauder proposed in Ref. [11] a formulation of coherent states for systems with continuous spectrum. They introduced a Hamiltonian $\hat{H} > 0$, with a non-degenerate continuous spectrum, thus

$$\hat{H}|\varepsilon\rangle = \omega\varepsilon|\varepsilon\rangle, \quad 0 < \varepsilon < \varepsilon_M \tag{14}$$

where $\{|\varepsilon\rangle\}$ stands for a basis of eigenstates, which we can generalize replacing suitably discrete parameters by continuous ones, sums by integrals and Kronecker by Dirac delta function [46]. In such a case, we can always chose a normalized basis of eigenvectors to rephrase Eqs. (7) and (8) in the following manner [46]

$$\langle\varepsilon|\varepsilon'\rangle = \delta(\varepsilon-\varepsilon'), \tag{15}$$

and

$$\int_0^{\varepsilon_M} |\varepsilon'\rangle\langle\varepsilon| = \hat{1}, \tag{16}$$

where $\varepsilon_M \leq \infty$ [11]. In the section 5 and here we use units in which $\hbar = 1$.

If we set $M(s) = e^{|z|^2/2}$ and $z = se^{-i\gamma\varepsilon}$ into coherent states (6), we find

$$|s,\gamma\rangle = M(s)^{-1} \int_0^{\varepsilon_M} d\varepsilon \frac{s^\varepsilon e^{-i\gamma\varepsilon}}{\sqrt{\rho(\varepsilon)}} |\varepsilon\rangle, \tag{17}$$

where $s > 0$. Since $\{|s,\gamma\rangle\}$ are orthonormals, the normalization factor $M(s)$ is given by

$$M(s)^2 = \int_0^{\varepsilon_M} d\varepsilon \frac{s^{2\varepsilon}}{\rho(\varepsilon)}, \tag{18}$$

for $M(s)^2 < \infty$.

Coherent states (17) must satisfy resolution of identity. In this case, it was introduced in Ref. [11] the following relation

$$\rho(\varepsilon) = \int_0^s ds' \, s'^{2\varepsilon}\sigma(s'), \tag{19}$$

where s' is a variable of integration with $0 \leq s' < s \leq \infty$. In addition, a non-negative weight function $\sigma(s') \geq 0$ was introduced in order to satisfy the second requirement. Then, the measure of integration takes the form [11]

$$d\tau(s,\gamma) = \sigma(s)M(s)^2 ds \frac{d\gamma}{2\pi}. \tag{20}$$

Gazeau and Klauder shown that resolution of unity is satisfied for systems with continuous spectrum in the present formulation of coherent states [11]. In Ref. [19] the authors have proposed a continuous appearance of Eq. (4), replacing the discrete form by the continuous version of variables, functions and operators involved in the formalism. Hence, we are ready to define the Husimi distribution for systems with continuous spectrum in the following manner:

$$\mu_Q(s,\gamma) = \frac{1}{Z} \int_0^{\varepsilon_M} d\varepsilon \, e^{-\beta\omega\varepsilon} |\langle s,\gamma|\varepsilon\rangle|^2, \tag{21}$$

where ε stands for a continuous parameter. The Husimi distribution is normalized according to

$$\int_0^\infty \int_{-\infty}^\infty d\tau(s,\gamma)\mu_Q(s,\gamma) = 1, \tag{22}$$

where the measure $d\tau(s,\gamma)$ is given by Eq. (20).

We see easily from Eq. (17) that, the projection of eingensatates of the Hamiltonian over coherent states, is given by

$$\langle s,\gamma|\varepsilon\rangle = M(s)^{-1}\frac{s^\varepsilon e^{-i\gamma\varepsilon}}{\sqrt{\rho(\varepsilon)}}, \tag{23}$$

where we have considered from Eq. (15) the orthogonality of the continuous states $\{|\varepsilon\rangle\}$. Introducing the above expression into Eq. (21) we finally arrive to [19]

$$\mu_Q(s) = \frac{M(s)^{-2}}{Z}\int_0^{\varepsilon_M} d\varepsilon\, \frac{e^{-\beta\omega\varepsilon}s^{2\varepsilon}}{\rho(\varepsilon)}, \tag{24}$$

where we have dropped out the dependence on γ. The continuous partition function obviously is [20]

$$Z = \int_0^{\varepsilon_M} d\varepsilon\, e^{-\beta\omega\varepsilon}. \tag{25}$$

It is important to note that Eq. (24) is consistently normalized in accordance with

$$\int_0^\infty d\tau(s)\mu_Q(s) = 1, \tag{26}$$

and in this case, the measure is $d\tau(s) = \sigma(s)M(s)^2\,ds$.

3. Landau diamagnetism: Charged particle in a uniform magnetic field

Diamagnetism was a problem firstly appointed by Landau who showed the discreteness of energy levels for a charged particle in a magnetic field [24]. By the observation of the diverse scenarios in the framework provided by the Landau diamagnetism we can study some relevant physical properties [25–27] as thermodynamic limit, role of boundaries, decoherence induced by the environment. The main motivation for several specialists work even today it is to make an accurate description of its theoretical and practical consequences.

In the past the appropriate partition function for this problem was calculated by Feldman and Kahn appealing to the concept of Glauber's coherent states as a set of basis states [28]. This formulation allows the use of classical concepts to describe electron orbits, even containing all quantum effects [28]. In a previous effort, this approach was used to obtain the Wehrl entropy [21, 22] and Fisher information [29] with the purpose of studying the thermodynamics of the Landau diamagnetism problem, namely, a free spinless charged particle in a uniform magnetic field [7]. In such contribution the authors focussed only in the transverse motion of a particle. For this reason, it was necessary to normalize the Husimi distribution in order to arrive to a consistent expression for semiclassical measures [7, 8, 32].

Certainly, because the relevant effects seem to come only from the transverse motion, several efforts are made to describe this problem in two dimensions [7, 8, 27, 28, 32-34]. Furthermore, since the discovery of interesting phenomena, as the quantum Hall effect, there has been much interest in understanding the dynamics of electrons confined to move in two dimensions in the presence of a magnetic field perpendicular to the motion plane [31]. The confinement is possible at the *interface* between two materials, typically a semiconductor and an insulator, where a quantum well that traps the particles is formed, forbidding their motion in the direction perpendicular to the interface plane at low energies.

However, we propose here to discuss this problem in the most complete form (three dimensions), some results related to the behavior of the Wehrl entropy. From the present line of reasoning, it is concluded that the two-dimensional formulation is sufficient unto itself to explain the problem whenever the length of the cylindrical geometry of the system is large enough. Nevertheless, as suggested before, electronic devices are based in interfaces. Thus, this fact theoretically imposes a natural lower temperature bound that emerges from the analysis when three dimensions are considered [18].

3.1. The model of one charged particle in a magnetic field

We enter the present application by revisiting the complete set of coherent states of a spinless charged particle in a uniform magnetic field. Consider the classical kinetic momentum

$$\vec{\pi} = \vec{p} + \frac{q}{c}\vec{A},$$ (27)

of a particle of charge q, mass m_q, and linear momentum $\vec{p}$, subject to the action of a vector potential $\vec{A}$. These are the essential ingredients of the well-known Landau model for diamagnetism: a spinless charged particle in a magnetic field B (we follow the presentation of Feldman et al. [28]). The Hamiltonian reads [28]

$$H = \frac{\vec{\pi} \cdot \vec{\pi}}{2m_q},$$ (28)

and the magnetic field is $\vec{B} = \vec{\nabla} \times \vec{A}$. The vector potential is chosen in the symmetric gauge as $\vec{A} = (-By/2, Bx/2, 0)$, which corresponds to a uniform magnetic field along the $z-$direction.

By using the quantum formulation of the step-ladder operators [28], one needs to define the step operators as follows [28]

$$\hat{\pi}_\pm = \hat{p}_x \pm i\hat{p}_y \pm \frac{i\hbar}{2\ell_{\mathrm{B}}^2}(\hat{x} \pm i\hat{y}),$$ (29)

where the length

$$\ell_{\mathrm{B}} = (\hbar c/qB)^{1/2}$$ (30)

is the classical radius of the ground-state Landau orbit [28]. Motion along the z-axis is free [28]. For the transverse motion, the Hamiltonian specializes to [28]

$$\hat{H}_t = \frac{\hat{\pi}_+ \hat{\pi}_-}{2m_q} + \frac{1}{2}\hbar\Omega\hat{1}, \tag{31}$$

where an important quantity characterizes the problem, namely,

$$\Omega = qB/m_q c, \tag{32}$$

the cyclotron frequency [33]. The eigenstates $|N,m\rangle$ are determined by two quantum numbers: N (associated to the energy) and m (to the $z-$ projection of the angular momentum). As a consequence, they are simultaneously eigenstates of both $\hat{H}_t$ and the angular momentum operator $\hat{L}_z$ [28], so that

$$\hat{H}_t|N,m\rangle = \left(N + \frac{1}{2}\right)\hbar\Omega|N,m\rangle = E_N|N,m\rangle \tag{33}$$

and

$$\hat{L}_z|N,m\rangle = m\hbar|N,m\rangle. \tag{34}$$

We note that the eigenvalues of $\hat{L}_z$ are not bounded by below (m takes the values $-\infty,\ldots,-1,0,1,\ldots,N$) [28]. This agrees with the fact that the energies $(N+1/2)\hbar\Omega$ are infinitely degenerate [33]. Such a fact diminishes the physical relevance of phase-space localization for estimation purposes, as we shall see below. Moreover, L_z is not an independent constant of the motion [33].

There exists a analogous formulation of an charged particle in a magnetic field by Kowalski that takes into account the geometry of a circle [30] (and for a comparison with the Feldman formulation see Ref.[8]), but at this point, we choose the Feldman formulation to work because the measure is easily defined and the normalization condition and other semiclassical measures are well described.

3.2. Husimi distribution and Wehrl entropy

We will start our present endeavor defining the Hamiltonian $\hat{H} = \hat{H}_t + \hat{H}_l$ for a particle of mass m_q and charge q in a magnetic field B, where $\hat{H}_t = \hbar\Omega(\hat{N}+1/2)$ describes the transverse motion, being Ω the cyclotron frequency as defined by the Eq. (32) and $\hat{N}$ the number operator. In addition, the Hamiltonian $\hat{H}_l = \hat{p}_z^2/2m_q$ represents a longitudinal one-dimensional free motion. After constructing a coherent state basis, a possible way to define the Husimi function η, for the complete motion, is given by

$$\eta(x,p_x;y,p_y;p_z) = \langle\alpha,\xi,k_z|\hat{\rho}|\alpha,\xi,k_z\rangle, \tag{35}$$

where $\hat{\rho}$ is the thermal density operator and the set $\{|\alpha,\xi,k_z\rangle\}$ represents the coherent states for the motion in three dimensions. Taking the direct product $|\alpha,\xi,k_z\rangle \equiv |\alpha,\xi\rangle \otimes |k_z\rangle$, the set $\{|\alpha,\xi\rangle\}$ corresponds to the coherent states of the transverse motion and $\{|k_z\rangle\}$ to the longitudinal motion. Therefore, the thermal density operator is given by

$$\hat{\rho} = \frac{1}{Z}e^{-\beta(\hat{H}_l+\hat{H}_t)}, \tag{36}$$

where $\beta = 1/k_B T$, k_B the Boltzmann constant and T the temperature. Besides, Z is the partition function for the particle total motion. If Z is separated in a similar way as other physical properties are separated, it is possible to assure that $Z = Z_l Z_t$, where Z_t is the contribution for the transverse motion and Z_l the contribution for the one-dimensional free motion. Thus, the Husimi function [1] is written as

$$\eta = \frac{e^{-\beta p_z^2/2m_q}}{Z_l Z_t} \sum_{n,m} e^{-\beta\hbar\Omega(n+1/2)} |\langle n,m|\alpha,\xi\rangle|^2. \tag{37}$$

where

$$Z_l = (L/h)(2\pi m_q k_B T)^{1/2} \quad \text{and} \tag{38}$$
$$Z_t = A m_q \Omega / (4\pi\hbar \sinh(\beta\hbar\Omega/2)), \tag{39}$$

being L the length of the cylinder, $A = \pi R^2$ the area for cylindrical geometry [28]. In addition, the matrix element $|\langle n,m|\alpha,\xi\rangle|^2$ represents the probability of finding the charged particle in the coherent state $|\alpha,\xi\rangle$ and we can find its expression as defined previously [34].

It should be noticed that the distribution η can be written as follows

$$\eta = \eta_l(p_z)\,\eta_t(\lambda,p_x,y,p_y), \tag{40}$$

where η has been separated as a function of two distributions, namely, $\eta_l = \eta_l(p_z)$ and $\eta_t = \eta_t(x,p_x;y,p_y)$. The dependence on the variable z has been missed due to the explicit form of the hamiltonian $\hat{H}_l$. Accordingly, after summing in Eq. (37) we find

$$\eta_l = \frac{e^{-\beta p_z^2/2m_q}}{Z_l}, \tag{41}$$
$$\eta_t = \frac{2\pi\hbar}{A m_q \Omega} (1 - e^{-\beta\hbar\Omega}) e^{-(1-e^{-\beta\hbar\Omega})|\alpha|^2/2\ell_B^2}, \tag{42}$$

where the length ℓ_B is defined by the Eq. (30). From expressions (41) and (42), we emphasize again that $\eta_l(p_z)$ describes the free motion of the particle in the magnetic field direction and $\eta_t(x,p_x;y,p_y)$ the Landau levels due to the circular motion in a transverse plane to the magnetic field, similar to the harmonic oscillator of Eq. (9) since $|z|^2 \to |\alpha|^2/2\ell_B^2$. Consequently Eqs. (40), (41) and (42) together contain the complete description of the system. We noticed both distributions are naturally normalized in a standard form, i.e.,

$$\int \frac{dz dp_z}{h} \eta_l(p_z) = 1, \tag{43}$$

and

$$\int \frac{d^2\alpha d^2\xi}{4\pi^2\ell_B^4} \eta_t(x,p_x;y,p_y) = 1. \tag{44}$$

In consequence, both Eqs. (41) and (42), under conditions (43) and (44), bring a promising way to get the exact form of the Wehrl entropy. Furthermore, using the additivity as the most basic property of the entropy, we can state $W_{total} = W_l + W_t$. Hence,

$$W_l = -\int \frac{dz dp_z}{h} \eta_l(p_z) \ln \eta_l(p_z), \tag{45}$$

$$W_t = -\int \frac{d^2\alpha d^2\xi}{4\pi^2 \ell_B^4} \eta_t(x,p_x;y,p_y) \ln \eta_t(x,p_x;y,p_y), \tag{46}$$

where, as before, the subindex l stands for the longitudinal motion and t the transverse.

After evaluating the respective integrals in Eqs. (45) and (46), it is feasible to identify the two particular entropies

$$W_l = \frac{1}{2} + \ln\left(\frac{L}{\lambda}\right), \tag{47}$$

$$W_t = 1 - \ln\left(1 - e^{-\beta\hbar\Omega}\right) + \ln(g), \tag{48}$$

where $\lambda = h/(2\pi m_q k_B T)^{1/2}$ is the mean thermal wavelength of the particle and $g = \mathcal{A}/2\pi\ell_B^2$ stands for the degeneracy of a Landau level [35]. Indeed, Eq. (47) coincides with the classical entropy for a free particle in one dimension. Eq. (48) is the Wehrl entropy for the transverse motion and possesses a form for the one close to the harmonic oscillator entropy given by the Eq. (10), with the exception of a term associated with the degeneracy.

3.3. Semiclassical behavior and consequences

Although the total Wehrl entropy is expressed simply as follows

$$W_{total} = \frac{3}{2} - \ln(1 - e^{-\beta\hbar\Omega}) + \ln(g) + \ln\left(\frac{L}{\lambda}\right), \tag{49}$$

we notice that some of its properties are directly derived from Eqs. (47) and (48). First, as we commented before, W_l coincides with the classical entropy for the free motion in one dimension. From this glance, we can add that W_l has to be nonnegative, $W_l \geq 0$ at all temperatures. This last condition imposes a minimum temperature, given by

$$T_0 = \frac{h^2}{2\pi m_q e k_B L^2}, \tag{50}$$

where $e = 2.718281828$. The standard behavior of W_l obligates the system to take high values of temperature, wherever the temperature T ought to be greater than T_0, in such case the conduct of the system is classical. This is equivalent to assert that, if $T/T_0 \geq 1$, the length of a thermal wave λ lower than the average of the spacing among particles and quantum considerations are not relevant [36]. In addition, T_0 only depends on the size of the system and does not depend on other external or internal physical parameters such as transverse area, external magnetic field, charge of the particle, etc. If the system is large then the minimum temperature is low. However, modern electronic systems has junctions where L is practically zero. In such case the required minimum temperature to make applicable our description is numerically high enough [39].

Nevertheless, the entropy associated with transverse motion satisfies $W_t \geq 1 + \ln(g)$ for all temperatures in the system of a particle in a magnetic field where the symmetry is polar, which is almost the Lieb condition for systems in one dimension [37] with an additional term associated with the degeneracy g. Roughly speaking, the transverse motion is bi-dimensional, but in the Landau approach the quantum motion of the particle in a magnetic field is reduced to a degenerate spectrum in one dimension. This degeneracy essentially recovers the physics of the missing dimension. Resuming the discussion of the behavior of the Wehrl entropy, it is not plausible to adventure any conclusion about the applicability of the present treatment because the Lieb condition is always satisfied. This is the main problem stems from the restricted vision presented in other contributions over this topic which only put its emphasis on the transverse motion [8, 28, 30] and represent the main difference from the vision obtained in that other contributions that discuss this topic. From the combined reasoning of both motions we conclude that the present description, this is the calculation of W_t, has sense when the imposition over the temperature is satisfied. Under T_0 the behavior is intrinsically anomalous and the present proposal is not applicable.

If we consider $k_B T \gg \hbar\Omega$, we can apply the first order of approximation as $\ln(g/(1 - e^{-\beta\hbar\Omega})) \approx \ln(\mathcal{A}T/T_0 L^2)$. Indeed, taking into account that the thermal wave length can be rewritten in terms of the temperature T_0 this way $\lambda = L(eT_0/T)^{1/2}$, the expression (49) after a bit of algebra reduces to

$$W_{\text{total}}^{(1)} \approx \frac{3}{2} \ln\left(\frac{T}{T_0}\right) + \ln\left(\frac{\mathcal{A}}{L^2}\right). \tag{51}$$

Considering that $\mathcal{V} = \mathcal{A}L$ in Eq. (51), the total Wehrl entropy can be expressed as follows

$$W_{\text{total}}^{(1)} = \frac{3}{2} + \ln\left(\frac{\mathcal{V}}{\lambda^3}\right). \tag{52}$$

This is a particular expression for the entropy of a free particle in three dimensions related to the motion of a charged particle into a region of the magnetic field making mention of some geometrical properties of the system.

In second order of approximation for high temperatures, considering the special condition $\mathcal{A} \sim L^2$, Wehrl entropy is expressed as follows

$$W_{\text{total}}^{(2)} \approx \frac{T_0}{T}g + \frac{3}{2} + \frac{3}{2} \ln\left(\frac{T}{T_0}\right) = \frac{T_0}{T}g + W_{\text{total}}^{(1)}. \tag{53}$$

As explained before, the Wehrl entropy takes values that are permitted by the Lieb condition, namely, $W \geq 1$. According to Eq. (53) the slope decreases as temperature increases. This fact illustrates why the disorder slowly increases as the magnetic field increases too. Consequently, at extremely high temperatures as expected, the slope of the present linear dependence tends to zero apparently taking a constant value close to the corresponding classical entropy of the free particle in three dimensions.

The lower bound of temperature is related to $T/T_0 \to 1^+$, because this approach does not consider temperature values under T_0. The total Wehrl entropy is reduced to logarithm behavior of the magnetic field.

To study what occurs close to zero temperature, in accordance with Eq. (50), we need to take systems with $L \to \infty$ and after this consideration the transverse entropy of Eq. (48) can be seen as follows

$$W_t^{T \to 0^+} = 1 + \ln(g). \tag{54}$$

As we discussed before, this Wehrl entropy is also a kind of harmonic oscillator entropy and the lower bound complies with being greater than a bound limiting value of the temperature, which has been suggested by Wehrl and shown by Lieb, $W \geq 1$ [37]. Starting from this condition it must arrive to the following inequality for the magnetic field

$$g \geq 1, \tag{55}$$

where $g = q\mathcal{A}B/hc$ also accounts for the ratio between the flux of the magnetic field $\mathcal{A}B$ and the quantum of the magnetic flux given by $hc/q = 4.14 \times 10^{-7}[gauss/cm^2]$ [17]. Then the inequality (55) adopts the form

$$B \geq \frac{1}{\mathcal{A}} \frac{hc}{q} = B_0. \tag{56}$$

Therefore, the quantity $B_0 = hc/\mathcal{A}q$ becomes a bound limiting field that represents the minimum value for the external magnetic field. To study what occurs close to zero magnetic field we need to take systems with $\mathcal{A} \to \infty$.

For finite values of $\mathcal{A}$ and B lower than B_0 is manifested the Haas-van Alphen effect, which describes oscillations in the magnetization because at temperatures low enough the particles will tend to occupy the lowest energy states. Whereas if the value of the magnetic field decreases a less number of particles can be in the lowest state due to degeneracy is directly proportional to B [35]. Then, the transverse Wehrl entropy W_t is well defined for values of the magnetic field over B_0, this is $B/B_0 \geq 1$ and/or $g \to 1^+$.

We can assert that this description of the system is not quantum, we say that it is semiclassical; for instance, it does not contain the Haas-van Alphen effect, the same condition marks the beginning of one description and the ending of the other.

Other relevant effect that emerges from the Landau quantization [38] is the quantum Hall effect [39] which is a quantum-mechanical version of the Hall effect [31], observed in two-dimensional electron systems subjected to low temperatures and strong magnetic fields. The degeneracy is given by [17]

$$\phi = \nu\phi_0, \tag{57}$$

where $\phi_0 = hc/q$ is the quantum of the magnetic flux. The factor ν is related to the "filling factor" that takes integer values ($\nu = 1, 2, 3, \dots$). The discovery of the fractional quantum Hall effect [32] extend these values to rational fractions ($\nu = 1/3, 1/5, 5/2, 12/5, \dots$). The integer quantum Hall effect is simply explained in terms of the conductivity quantization $\sigma = \nu q^2/h$. However, the fractional quantum Hall effect relies on other phenomena related to interactions. Consistently, we see that the degeneracy is equal to v, which must be greater than 1 due to the inequality (55) obtaining an infinite family of Wehrl entropies

$$W_t = 1 - \ln(1 - e^{-\beta\hbar\Omega}) + \ln v. \tag{58}$$

Again, Eq. (55) provides the limiting value of v and, as before, the transverse entropy always satisfies the Lieb bound for all temperatures and large enough systems when the quantum Hall effect is manifested at least for the integer quantum Hall effect. Conversely, fractional values of v less than 1 are left out the present approach.

4. Description of the molecular rotation: Rigid rotator

The rigid rotator is a system of a single particle whose quantum spectrum of energy is exactly known. Therefore, the study of typical thermodynamic properties can be analytically derived [40]. Applications lead to the treatment of important aspects of molecular systems [41] and several applications to materials [42].

4.1. Linear rigid rotator

We start the present study by exploring a simple model, the linear rigid rotator, based on the excellent discussion concerning the coherent states for angular momenta given in Ref. [43]. The Hamiltonian of the linear rigid rotator is [20]

$$\hat{H} = \frac{\hat{L}^2}{2I_{xy}}, \tag{59}$$

where $\hat{L}^2 = \hat{L}_x^2 + \hat{L}_y^2$ is the angular momentum operator and I_x and I_y are the associated moments of inertia. We have assumed that $I_{xy} \equiv I_x = I_y$. Calling $|IK\rangle$ the set of H-eigenstates, we recall that they verify the relations

$$\begin{aligned}
\hat{L}^2|IK\rangle &= I(I+1)\hbar^2|IK\rangle \\
\hat{L}_z|IK\rangle &= K\hbar|IK\rangle,
\end{aligned} \tag{60}$$

with $I = 0, 1, 2 \ldots$, for $-I \leq K \leq I$, the eigenstates' energy spectrum being given by

$$\varepsilon_I = \frac{I(I+1)\hbar^2}{2I_{xy}}. \tag{61}$$

Coherent states are constructed in Ref. [44, 45] for the lineal rigid rotator, using Schwinger's oscillator model of angular momentum, in the fashion

$$|IK\rangle = \frac{(\hat{a}_+^\dagger)^{I+K}(\hat{a}_-^\dagger)^{I-K}}{\sqrt{(I+K)!(I-K)!}}|0\rangle, \tag{62}$$

with $\hat{a}_+$, $\hat{a}_-$ the pertinent creation and annihilation operators, respectively, and $|0\rangle \equiv |0,0\rangle$ the vacuum state. The states $|IK\rangle$ are orthogonal and satisfy the closure relation, i.e.,

$$\langle I'K'|IK\rangle = \delta_{I',I}\delta_{K',K},\tag{63}$$

$$\sum_{I=0}^{\infty}\sum_{K=-I}^{I}|IK\rangle\langle IK| = \hat{1}.\tag{64}$$

Since we deal with two degrees of freedom the ensuing coherent states are of the tensor product form (involving $|z_1\rangle$ and $|z_2\rangle$) [43, 46]

$$|z_1 z_2\rangle = |z_1\rangle \otimes |z_2\rangle,\tag{65}$$

where

$$\hat{a}_+|z_1 z_2\rangle = z_1|z_1 z_2\rangle,\tag{66}$$

$$\hat{a}_-|z_1 z_2\rangle = z_2|z_1 z_2\rangle.\tag{67}$$

Therefore, the coherent state $|z_1 z_2\rangle$ writes [43]

$$|z_1 z_2\rangle = e^{-\frac{|z|^2}{2}} e^{z_1 \hat{a}_+^{\dagger}} e^{z_2 \hat{a}_-^{\dagger}}|0\rangle,\tag{68}$$

with

$$|z_1\rangle = e^{-\frac{|z_1|^2}{2}} e^{z_1 \hat{a}_+^{\dagger}}|0\rangle,\tag{69}$$

$$|z_2\rangle = e^{-\frac{|z_2|^2}{2}} e^{z_2 \hat{a}_-^{\dagger}}|0\rangle.\tag{70}$$

We have introduced the convenient notation

$$|z|^2 = |z_1|^2 + |z_2|^2.\tag{71}$$

Using Eqs. (62) and (68) we easily calculate $|z_1 z_2\rangle$ and, after a bit of algebra, find

$$|z_1 z_2\rangle = e^{-\frac{|z|^2}{2}}\sum_{n_+,n_-}\frac{z_1^{n_+}}{\sqrt{n_+!}}\frac{z_2^{n_-}}{\sqrt{n_-!}}|IK\rangle\tag{72}$$

where $n_+ = I + K$ and $n_- = I - K$. Therefore, the probability of observing the state $|IK\rangle$ in the coherent state $|z_1 z_2\rangle$ is of the form

$$|\langle IK|z_1 z_2\rangle|^2 = e^{-|z|^2} \frac{|z_1|^{2n_+}}{n_+!} \frac{|z_2|^{2n_-}}{n_-!}. \tag{73}$$

The present coherent states satisfy resolution of unity

$$\int \frac{d^2 z_1}{\pi} \frac{d^2 z_2}{\pi} |z_1 z_2\rangle\langle z_1 z_2| = 1. \tag{74}$$

Furthermore, z_1 and z_2 are continuous variables.

Following the procedure developed by Anderson *et al.* [4], we can readily calculate the pertinent Husimi distribution [1]. For our system this is defined, from Eq. (4), as

$$\mu(z_1, z_2) = \langle z_1, z_2|\hat{\rho}|z_1, z_2\rangle, \tag{75}$$

where the density operator is

$$\hat{\rho} = Z_{2D}^{-1} \exp\left(\beta \hat{H}\right). \tag{76}$$

The concomitant rotational partition function Z_{2D} is given in Ref. [20]

$$Z_{2D} = \sum_{I=0}^{\infty} (2I+1) e^{-I(I+1)\frac{\Theta}{T}}, \tag{77}$$

with $\Theta = \hbar^2/(2I_{xy}k_B)$. Remark that in the present context, speaking of the "trace operation" entails performing the sum $\text{Tr} \equiv \sum_{I=0}^{\infty}\sum_{K=-I}^{I}$. Inserting now the closure relation into Eq. (75), and using Eq. (73), we finally get our Husimi distribution in the fashion

$$\mu(z_1, z_2) = e^{-|z|^2} \frac{\sum_{I=0}^{\infty} \frac{|z|^{4I}}{(2I)!} e^{-I(I+1)\frac{\Theta}{T}}}{\sum_{I=0}^{\infty} (2I+1) e^{-I(I+1)\frac{\Theta}{T}}}. \tag{78}$$

It is easy to show that this distribution is normalized to unity

$$\int \frac{d^2 z_1}{\pi} \frac{d^2 z_2}{\pi} \mu(z_1, z_2) = 1, \tag{79}$$

where z_1 and z_2 are given by Eqs. (66), (67), and (71). Note that we must deal with the binomial expression $(|z_1|^2 + |z_2|^2)^{4I}$ firstly and then integrate over the whole complex plane (in two dimensions) in order to verify the normalization condition (79). The differential element of area in the $z_1(z_2)$ plane is $d^2 z_1 = dxdp_x/2\hbar$ ($d^2 z_2 = dydp_y/2\hbar$) [13]. Moreover, we have the phase-space relationships

$$|z_1|^2 = \frac{1}{4}\left(\frac{x^2}{\sigma_x^2} + \frac{p_x^2}{\sigma_{p_x}^2}\right), \tag{80}$$

$$|z_2|^2 = \frac{1}{4}\left(\frac{y^2}{\sigma_y^2} + \frac{p_y^2}{\sigma_{p_y}^2}\right), \tag{81}$$

where $\sigma_x \equiv \sigma_y = \sqrt{\hbar/2m\omega}$ and $\sigma_{p_x} \equiv \sigma_{p_y} = \sqrt{m\omega\hbar/2}$.

The profile of the Husimi function is similar to that of a Gaussian distribution.

The Wehrl entropy is a semiclassical measure of localization [21] (so is Fisher's one [5] as well). Indeed, Wehrl's measure is simply a logarithmic Shannon measure built up with Husimi distributions. For the present bi-dimensional model this entropy reads

$$\mathcal{W} = -\int \frac{d^2 z_1}{\pi}\frac{d^2 z_2}{\pi}\,\mu(z_1, z_2)\,\ln\mu(z_1, z_2), \tag{82}$$

where $\mu(z_1, z_2)$ is given by Eq. (78).

4.2. Rigid rotator in three dimensions

In the present section we consider a more general problem, the model of the rigid rotator in three dimensions, whose Hamiltonian writes [47]

$$\hat{H} = \frac{\hat{L}_x^2}{2I_x} + \frac{\hat{L}_y^2}{2I_y} + \frac{\hat{L}_z^2}{2I_z}, \tag{83}$$

where I_x, I_y, and I_z are the associated moments of inertia. A complete set of rotator eigenstates is $\{|IMK\rangle\}$. The following relations apply

$$\begin{aligned}
\hat{L}^2|IMK\rangle &= I(I+1)\hbar^2|IMK\rangle \\
\hat{L}_z|IMK\rangle &= K\hbar|IMK\rangle \\
\hat{J}_z|IMK\rangle &= M\hbar|IMK\rangle,
\end{aligned} \tag{84}$$

where $I = 0,\ldots,\infty, -I \leq K \leq I$, and $-I \leq M \leq I$. The states $|IMK\rangle$ satisfy orthogonality and closure relations [47]

$$\langle I'M'K'|IMK\rangle = \delta_{I',I}\delta_{M',M}\delta_{K',K} \tag{85}$$

$$\sum_{I=0}^{\infty} \sum_{M=-I}^{I} \sum_{K=-I}^{I} |IMK\rangle\langle IMK| = \hat{1}. \tag{86}$$

If we take $\hat{L}^2 = \hat{L}_x^2 + \hat{L}_y^2 + \hat{L}_z^2$ and assume axial symmetry, i.e., $I_{xy} \equiv I_x = I_y$, we can recast the Hamiltonian as

$$\hat{H} = \frac{1}{2I_{xy}} \left[\hat{L}^2 + \left(\frac{I_{xy}}{I_z} - 1 \right) \hat{L}_z^2 \right], \tag{87}$$

where $\hat{L}^2$ is the angular momentum operator and $\hat{L}_z$ is its projection on the rotation axis z. The concomitant spectrum of energy becomes

$$\varepsilon_{I,K} = \frac{\hbar^2}{2I_{xy}} \left[I(I+1) + \left(\frac{I_{xy}}{I_z} - 1 \right) K^2 \right], \tag{88}$$

where $I = 0, 1, 2, \cdots$ and it represents the eigenvalue of the angular momentum operator $\hat{L}^2$, the numbers $m = -I, \cdots, -1, 0, 1, \cdots, I$ stand for the projections on the intrinsic rotation axis of the rotator. All states exhibit a $(2I+1)$–degeneracy. The parameters $I_x = I_y \equiv I_{xy}$ and I_z are the inertia momenta. Different "geometrical" instances are characterized through the I_{xy}/I_z–ratio. For example, the value $I_{xy}/I_z = 1$ corresponds to the spherical rotator. Limiting cases can also be considered. This is, $I_{xy}/I_z = 1/2$ and $I_{xy}/I_z \to \infty$, that correspond to the extremely oblate- and prolate cases, respectively.

4.2.1. Coherent states for the rigid rotator in three dimensions

In order to obtain the Husimi distribution for this problem we need first of all to have the associated coherent states. Morales et al. have constructed them in Ref. [47] and discussed their mathematical foundations. First, they introduced the auxiliary quantity

$$X_{I,M,K} = \sqrt{I!(I+M)!(I-M)!(I+K)!(I-K)!}, \tag{89}$$

and then write [47]

$$|z_1 z_2 z_3\rangle = e^{-\frac{|u|^2}{2}} \sum_{IMK} \frac{[(2I)!]^2 z_1^{(I+M)} z_2^I z_3^{(I+K)}}{X_{I,M,K}} |IMK\rangle, \tag{90}$$

where the following supplementary variable were introduced by Morales et al. in Ref. [47]

$$|u|^2 = |z_2|^2 (1 + |z_1|^2)^2 (1 + |z_3|^2)^2. \tag{91}$$

All coherent states share at least two requirements. Continuity of labeling and resolution of unity. In relation to the last property we add

$$\int d\Gamma |z_1 z_2 z_3\rangle\langle z_1 z_2 z_3| = 1 \tag{92}$$

where $d\Gamma$ is the measure of integration given by [47]

$$d\Gamma = d\tau \left\{ 4[(1+|z_1|^2)(1+|z_3|^2)]^4 |z_2|^4 - 8[(1+|z_1|^2)(1+|z_3|^2)]^2 |z_2|^2 + 1 \right\} \tag{93}$$

with

$$d\tau = \frac{d^2 z_1}{\pi} \frac{d^2 z_2}{\pi} \frac{d^2 z_3}{\pi}, \tag{94}$$

and, of course, in this case we have three degrees of freedom. The present formulation satisfy the weaker version of the second requirement, because the measure is defined non positive [47].

4.2.2. Husimi function, Wehrl entropy

Using now Eq. (90) we find

$$|\langle IMK|z_1 z_2 z_3 \rangle|^2 = \frac{e^{-|u|^2}}{X_{I,M,K}^2} [(2I)!!]^2 |z_1|^{2(I+M)} |z_2|^{2I} |z_3|^{2(I+K)} \tag{95}$$

and determine that, in this case, the rotational partition function reads

$$Z_{3D} = \sum_{I=0}^{\infty} \sum_{K=-I}^{I} \sum_{M=-I}^{I} e^{-\beta \varepsilon_{I,K}}, \tag{96}$$

i.e.,

$$Z_{3D} = \sum_{I=0}^{\infty} (2I+1) e^{-I(I+1)\frac{\Theta}{T}} \sum_{K=-I}^{I} e^{-\left(\frac{I_{xy}}{I_z}-1\right)K^2 \frac{\Theta}{T}}. \tag{97}$$

Remark that if we take the "extremely prolate" limiting case $I_{xy}/I_z \to \infty$ just one term that survives in the right sum of the right side in Eq. (97), that for $K = 0$, while all terms for $K \neq 0$ vanish. In this special instance case Z_{2D} is recovered from Z_{3D}. The pertinent Husimi distribution becomes

$$\mu(z_1, z_2, z_3) = \frac{e^{-|u|^2}}{Z_{3D}} \sum_{I=0}^{\infty} \frac{(2I)!}{I!} |v|^{2I} e^{-I(I+1)\frac{\Theta}{T}} \times g(I), \tag{98}$$

where

$$g(I) = \sum_{K=-I}^{I} \frac{|z_3|^{2(I+K)}}{(I+K)!(I-K)!} e^{-\left(\frac{I_{xy}}{I_z}-1\right)K^2 \frac{\Theta}{T}}, \tag{99}$$

with

$$|v|^2 = (1+|z_1|^2)^2 |z_2|^2, \tag{100}$$

$$|u|^2 = |v|^2(1 + |z_3|^2)^2. \tag{101}$$

We can easily verify that $\mu(z_1, z_2, z_3)$ is normalized in the fashion

$$\int d\Gamma \, \mu(z_1, z_2, z_3) = 1, \tag{102}$$

We compute now (i) the Wehrl entropy in the form

$$\mathcal{W} = \int d\Gamma \, \mu(z_1, z_2, z_3) \, \ln \mu(z_1, z_2, z_3). \tag{103}$$

In the special instance $I_{xy}/I_z = 1$, that corresponds to the spherical rotator, we explicitly obtain

$$\mu(z_1, z_2, z_3) = e^{-|u|^2} \frac{\sum_{l=0}^{\infty} \frac{|u|^{2l}}{l!} e^{-l(l+1)\frac{\Theta}{T}}}{\sum_{l=0}^{\infty} (2l+1)^2 e^{-l(l+1)\frac{\Theta}{T}}}. \tag{104}$$

Having the Husimi functions the Wehrl entropy is straightforwardly computed.

In order to emphasize some special cases associated to possible applications we consider several possibilities.

1. The spherical rotator $I_{xy} = I_x = I_y = I_z$, thus $I_{xy}/I_z = 1$ (e.g. CH_4).
2. The oblate rotator $I_{xy} = I_x = I_y < I_z$, specifically $1/2 \leq I_{xy}/I_z < 1$ (e.g. C_6H_6).
3. The prolate rotator $I_{xy} = I_x = I_y > I_z$, which corresponds to $I_{xy}/I_z > 1$ (e.g. PCl_5).
4. The extremely prolate rotator is equivalent to the linear case (all diatomic molecules, $I_z = 0$, this is $I_{xy}/I_z \to \infty$ (e.g. CO_2, C_2H_2).

5. Husimi distribution for systems with continuous spectrum

In this section we propose a procedure to generalize the Husimi distribution to systems with continuous spectrum. We start extending the concept of coherent states for systems with discrete spectrum to systems with continuous one. In the present section, we see the Husimi distribution as a representation of the density operator in terms of a basis of coherent states. We specially discuss the problem of the continuous harmonic oscillator [20].

5.1. The exponential weight function: Harmonic oscillator

From the $\rho(\varepsilon)$ definition expressed in Eq. (19), we can take a non-negative weight function like $\sigma(s') = \exp(-s')$. However, this choice is not fully arbitrary, because it relies on , at least, two reasons: 1) it is related to the harmonic oscillator and, 2) it is a useful function that permits exactly to solve the integral (19). The latter reason allows to express such integral in the following way

$$\rho(\varepsilon) = \int_0^s ds' s'^{2\varepsilon} \exp(-s'),$$

$$= e^{-s/2} \frac{s^{\varepsilon}}{2\varepsilon+1} \mathcal{M}(\varepsilon, \varepsilon+1/2, s) \tag{105}$$

where $\mathcal{M}(a,b,x)$ is the Whittaker function [48]. Besides, in relation to the first reason, when we consider $\varepsilon = n$, where n is integer, in the limit $s \to \infty$; the Eq. (105) drops into the known quantum result for the harmonic oscillator, $\rho(n) = n!$ [11].

Moreover, the measure in phase space can be explicitly expressed from Eq. (20) as follows

$$d\tau(s) = ds\, e^{-s/2} \int_0^{\varepsilon_F} d\varepsilon \frac{(2\varepsilon+1)s^{\varepsilon}}{\mathcal{M}(\varepsilon, \varepsilon+1/2, s)}. \tag{106}$$

Although obtaining this explicit form of the measure, a most general expression for the integral of Eq. (106) strongly depends on the particular spectrum of the system. In the present case, a spectrum like $\varepsilon \propto \omega$, the harmonic oscillator in the continuous limit, is considered.

5.2. $s \to 0$ approximation for the Husimi distribution

In order to know the shape of the Husimi distribution in $s = 0$, we need to calculate some important quantities. First, we evaluate $\rho(\varepsilon)$ given by Eq. (105) expanding the exponential which appears inside the integral, as follows

$$\rho(\varepsilon) \approx \lim_{s \to 0} \int_0^s ds' s'^{2\varepsilon}(1 - s' + \cdots), \tag{107}$$

$$\approx \frac{s^{2\varepsilon+1}}{2\varepsilon+1}\left(1 - \frac{2\varepsilon+1}{2\varepsilon+2}s + \cdots\right). \tag{108}$$

But, we are interested in evaluating the inverse of $\rho(\varepsilon)$, therefore

$$\frac{1}{\rho(\varepsilon)} \approx \frac{2\varepsilon+1}{s^{2\varepsilon+1}}\left(1 + \frac{2\varepsilon+1}{2\varepsilon+2}s + \cdots\right). \tag{109}$$

Second, we show easily that, in the limit $s \to 0$, the Husimi distribution is given by

$$\mu_Q(0) = \frac{1}{Z}\frac{\int_0^{\varepsilon_M} d\varepsilon\,(2\varepsilon+1)e^{-\beta\omega\varepsilon}}{\int_0^{\varepsilon_M} d\varepsilon\,(2\varepsilon+1)}. \tag{110}$$

Now, after integrating Eq. (25) the partition function is expressed as follows

$$Z = \frac{1 - \exp(-\beta\omega\varepsilon_M)}{\beta\omega}. \tag{111}$$

Then, the substitution of the Eq. (111) into (110) leads us to the appearance

$$\mu_Q(0) = \frac{(2e^{-\beta\omega\varepsilon_M}\beta\omega\varepsilon_M + 2e^{-\beta\omega\varepsilon_M} + e^{-\beta\omega\varepsilon_M}\beta\omega - 2 - \beta\omega)}{\beta\omega\varepsilon_M(e^{-\beta\omega\varepsilon_M} - 1)(\varepsilon_M + 1)}. \tag{112}$$

In the high temperature limit, this becomes

$$\mu_Q(0) \approx \frac{1}{\varepsilon_M} - \frac{\beta\omega\varepsilon_M}{6(\varepsilon_M + 1)}. \tag{113}$$

If we take into account a kind of particles filling a band in the lowest continuous levels of energy (for instance, $\varepsilon_M \to 1$), we find $\mu_Q(0) = 1 - \beta\omega/12$.

5.3. Asymptotic behavior of the Husimi function

In this part of the work, we are considering a particular range for ε; *i.e.*, $0 \leq \varepsilon \leq \varepsilon_M = 1$ and we study the asymptotic behavior of the Husimi distribution. This trend might be obtained from the limiting case of the Whittaker function [48] defined for $s \to \infty$, as follows:

$$\lim_{s\to\infty} \frac{e^{-s/2}s^{\varepsilon}\mathcal{M}(\varepsilon, \varepsilon + 1/2, s)}{2\varepsilon + 1} = \Gamma(2\varepsilon + 1). \tag{114}$$

If we replace this result into Eq. (124) we obtain

$$M(s)^2 = e^{s/2}\int_0^{\varepsilon_M} d\varepsilon \frac{s^{2\varepsilon}}{\Gamma(2\varepsilon + 1)}, \tag{115}$$

and, from Eq. (24) we write

$$\mu_Q(s) = \frac{M(s)^{-2}}{Z}e^{s/2}\int_0^{\varepsilon_M} d\varepsilon \frac{e^{-\omega\beta\varepsilon}s^{2\varepsilon}}{\Gamma(2\varepsilon + 1)}. \tag{116}$$

Now, we follow expanding to third order the inverse of the gamma function, $1/\Gamma(2\varepsilon + 1)$, around its maximum [48]

$$\frac{1}{\Gamma(2\varepsilon + 1)} \approx \sum_{n=0}^{3} A_n\varepsilon^n, \tag{117}$$

where $A_0 = .9963530195$, $A_1 = 1.221909147$, $A_2 = -3.108524622$, and $A_3 = 1.333217620$.

From Eq. (115), we derive a approximate result for $M(s)^2$, which is given by

$$M(s)^2 = e^{s/2}\frac{s}{2}\sum_{n=0}^{3} A_n \frac{\mathcal{M}\left(\frac{n}{2}, \frac{n}{2} + \frac{1}{2}, -2\ln(s)\right)}{(n+1)(-2\ln(s))^{1+n/2}}, \tag{118}$$

and combining all above expressions, we have finally found an expression to third order of approximation for Husimi distribution given by

$$\mu_Q(s) = \frac{M(s)^{-2}}{Z} e^{s/2 - \beta\omega/2} \frac{s}{2} \sum_{n=0}^{3} A_n \frac{\mathcal{M}\left(\frac{n}{2}, \frac{n}{2} + \frac{1}{2}, \beta\omega - 2\ln(s)\right)}{(n+1)(\beta\omega - 2\ln(s))^{1+n/2}}, \tag{119}$$

where $\mathcal{M}(a,b,c)$ is again the Whittaker function [48].

In the high temperature approximation, Eq. (119) is given by

$$\mu_Q(s) \approx \beta\omega \frac{\exp(-\beta\omega/2)}{1 - \exp(-\beta\omega)} \approx \exp(-\beta\omega/2). \tag{120}$$

The present result does not depend on the values of the parameter s. Furthermore, this approximation is valid whenever $0 \leq \varepsilon \leq 1$. We notice that the asymptotic trend of the Husimi distribution approaches to the Boltzmann weight in the ground state of the harmonic oscillator.

5.4. Some applications and consequences

In Ref. [11], the mean value of energy is obtained from the expectation value of the classical Hamiltonian $\mathcal{H}$ in a coherent state as follows $\mathcal{H}(s) = \langle s, \gamma | \mathcal{H} | s, \gamma \rangle$, therefore they arrive to the relation $\mathcal{H}(s) = s \partial \ln M(s)/\partial s$.

However, it is our interest here to calculate the mean value of energy in a different way, integrating in the variable s with $\mu_Q(s)$ as a weigh function. Hence, we have

$$\langle \mathcal{H} \rangle = \int d\tau(s) \mu_Q(s) \mathcal{H}(s), \tag{121}$$

where $\mathcal{H}$, expressed in terms of the variable s, denotes the classical Hamiltonian of the system. Inserting the Husimi distribution (24) into Eq. (121) and making use the relation (19) we finally get

$$\langle \mathcal{H} \rangle = \frac{1}{Z} \int_0^{\varepsilon_M} d\varepsilon\, e^{-\beta\varepsilon} \mathcal{H}(\varepsilon), \tag{122}$$

that is the classical mean energy [20]. We emphasize that the Husimi distribution, for a system with continuous spectrum, conduces in a natural way to the classical mean value of energy. Obviously, this is not true when the spectrum is discrete.

An extra motivation consists in extending the formulation of coherent states to systems with continuum spectrum considering its explicit form; for instance, we can take a spectrum whose appearance is $E = A\varepsilon^\nu$, where A and ν are constant. The values $\nu = \pm 1$ and $\nu = 2$ might define the continuous limit of three remarkable cases in physics. Certainly, in a general study other values of the parameter ν may be conveniently considered as an interesting analytical extension. Thus, for $\nu = 1$ and $A = \omega$ we have the continuous limit of a particle in a harmonic potential; this case is being in detail discussed in the present work. For $\nu = 2$ we have the continuous limit of a particle in a box. For $\nu = -1$ we have the continuous limit of a particle in a Coulomb potential. Therefore, it is necessary to introduce a density of states $g(E)$ in the formulation of continuous coherent states (17) and immediately get the following modification

$$|s, \gamma\rangle = M(s)^{-1} \int_0^{E_M} dE\, g(E) \frac{s^{E/A} e^{-i\gamma E/A}}{\sqrt{\rho(E)}} |\varepsilon\rangle, \tag{123}$$

where the function $M(s)$

$$M(s)^2 = \int_0^{E_M} dE\, g(E)^2 \frac{s^{2E/A}}{\rho(E)} \tag{124}$$

represents the normalization factor.

6. Final remarks

We have included in the current work some motivational elements to develop possible future applications to information theory and condensed matter. We have focused attention primarily upon Husimi distribution and its analytical results, beyond the numerical, graphical, or approximate calculations. A semiclassical description undertaking can be tackled, (i) trying to estimate phase-space location via measures as Fisher information and (ii) evaluating the semiclassical Wehrl entropy. A crucial point, in such an estimation, is to define the Husimi distribution in a convenient set of coherent states. Hence, we introduce a formal view of general requirements for formulations of coherent states in the context of the Gazeau and Klauder formalism for the harmonic oscillator – we have included some mathematical details in order to make it easy to follow and instructive in courses of quantum mechanics for graduates– we show some practical elements to apply the present formalisms to specific calculations of semi-classical measures.

By using a suitable formulation of coherent states in every case, we show explicitly the form of the Husimi distribution for i) a spinless charged particle in a uniform magnetic field (Landau diamagnetism), (ii) the linear and the three dimensional rotator (molecular rotation) and (iii) a case of the limiting harmonic oscillator (continuous spectrum).

In addition, we can calculate the probability by projecting the states over the coherent states as a function of a variable related to the coherent states. We see that the localization of probability, in the phase space decreases as temperature increases. Also, as always, the localization of the Husimi distribution in the phase space decreases as temperature increases. The present derivation of Husimi distributions is based on the evaluation of the mean value of the density operator in the basis of a single-particle coherent state. While the Husimi function takes into account collective and environmental effects, the coherent states are independent-particle states. Thus, if the Husimi distribution is delocalized, we need many wave packets (independent-particle states) to represent the state. Furthermore, the thermodynamics of particles in systems, which come from environmentally induced effects, does not depend on the formulation of the coherent states. In this manner, we expect this behavior to become general.

In conclusion, quantal distributions in the phase space, such as the Husimi distribution, have long been recognized as powerful tools for studying the quantum-classical correspondence and semi-classical aspects of quantum mechanics, since they provide a phase-space picture of the density matrix. We aknowledge partial financial support by FONDECYT 1110827.

Author details

Sergio Curilef[1] and Flavia Pennini[1,2]

1 Departamento de Física, Universidad Católica del Norte, Antofagasta, Chile
2 Instituto de Física La Plata–CCT-CONICET, Fac. de Ciencias Exactas, Universidad Nacional de La Plata, La Plata, Argentina

References

[1] K. Husimi *Proceedings of the Physico-Mathematical Society of Japan* 22 (1940) 264 .

[2] K. Takahashi and N. Saitô, *Physical Review Letters* 55 (1985) 645.

[3] M.C. Gutzwiller, *Chaos in Classical and Quantum Mechanics*, (Springer-Verlag, New York, 1990).

[4] A. Anderson and J.J. Halliwell, *Physical Review D* 48 (1993) 2753.

[5] F. Pennini and A. Plastino, *Physical Review* E 69 (2004) 057101.

[6] S. Curilef, F. Pennini, A. Plastino and G.L. Ferri, *J. Phys. A: Math. Theor.* 40 (2007) 5127.

[7] S. Curilef, F. Pennini and A. Plastino, *Physical Review* B 71 (2005) 024420.

[8] D. Herrera, A.M. Valencia, F. Pennini and S. Curilef, *European Journal of Physics* 29 (2008) 439.

[9] M. Janssen, *Fluctuations and Localization in Mesoscopic Electron Systems*, World Scientific Lecture Notes in Physics Vol. 64., 2001

[10] W.P. Scheleich, *Quantum Optics in phase space*, (Wiley VCH-Verlag, Berlin, 2001).

[11] J.P. Gazeau and J.R. Klauder, *Journal of Physics A: Math Gen.* 32 (1999) 123.

[12] E. Schrödinger, *Naturwissenschaften* 14 (1926) 664.

[13] R.J. Glauber, *Physical Review Letters* 10 (1963) 84.

[14] R.J. Glauber, *Physical Review* 131 (1963) 2766.

[15] V.V. Dodonov and V.I. Man'ko, *Theory of nonclassical states of light*, (Taylor & Francis Group, London and New York, 2003).

[16] J.R. Klauder, *J. Phys. A* 29 (1996) L293.

[17] F. Olivares, F. Pennini, S. Curilef, *Physical Review* E 81 (2010) 041134.

[18] S. Curilef, F. Pennini, A. Plastino, G.L. Ferri, *Journal of Physics: Conference Series* 134 (2008) 012029.

[19] F. Pennini, S. Curilef, *Communications in Theoretical physics* 53 (2010) 535.

[20] R.K. Pathria, *Statistical Mechanics*, (Pergamon Press, Exeter, 1993).

[21] A. Wehrl, *Reviews of Modern Physics* 50 (1978) 221.

[22] A. Wehrl, *Reports on Mathematical Physics* 16 (1979) 353.

[23] M. Thaik and A. Inomata, *Journal of Physics. A: Math Gen.* 38 (2005) 1767.

[24] L.D. Landau, *Z. Physik* 64 (1930) 629.

[25] S. Dattagupta, A.M. Jayannvar, N. Kumar, *Current Science* 80 (2001) 861.

[26] J. Kumar, P.A. Sreeram, S. Dattagupta, *Physical Review* E 79 (2009) 021130.

[27] A. M. Jayannavar, M. Sahoo, *Physical Review* E 75 (2007) 032102.

[28] A. Feldman, A.H. Kahn, *Physical Review* B 1 (1970) 4584.

[29] B.R. Frieden, (Cambridge University Press, Crambridge, England, 1998).

[30] K. Kowalski, J. Rembieliński, *Journal of Physics A: Math. Gen.* 38 (2005) 8247.

[31] K. v. Klitzing, G. Dorda, M. Pepper, *Physical Review Letters* 45 (1980) 494.

[32] D.C. Tsui, II.L. Stormer, A.C. Gossard, *Physical Review Letters* 48 (1982) 1559.

[33] M.H. Johnson and B. A. Lippmann, *Physical Review* 76 (1949) 828.

[34] See Eq. (16) in Ref.[7].

[35] K. Huang, *Statistical Mechanics*, Wiley, New York, Second edition (1963).

[36] L.E. Reichl, *A Modern course in statistical physics*, Wiley, New York, Second edition, (1998).

[37] E.H. Lieb, *Communications in Mathematical Physics* 62 (1978) 35.

[38] F. Lado, *Physics Letters* A 312 (2003) 101.

[39] T. Chakraborty, P. Pietiläinen, *The Quantum Hall Effects*, Springer-Verlag, Berlin, (1995).

[40] N. Ullah, *Physical Review* E 49 (1994) 1743.

[41] F.J. Arranz, F. Borondo, R.M. Benito, *Physical Review* E 54 (1996) 2458.

[42] J.T. Titantah, M.N. Hounkonnou, *Journal of Physics A: Math. Gen.* 32 (1999) 897; 30 (1997) 6347; 30 (1997) 6327; 28 (1995) 6345.

[43] M.M. Nieto, *Physical Review* D 22 (1980) 391.

[44] J.J. Sakurai, Modern Quantum Mechanicsed S.F. Tuan (Reading, MA: Addison-Wesley, 1994) p 217.

[45] J. Schwinger, In Quantum Theory of Angular Momentumed, L. C. Biedenharn and H van Dam (New York: Academic, 1965) p 229.

[46] C. Cohen-Tannoudji, B. Diu, F. Laloe, Quantum Mechanics vol.1 John Wiley & Sons, (1977).

[47] J.A. Morales, E. Deumens and Y. Öhrn, *Journal of Mathematical Physics* 40 (1999) 776.

[48] M. Abramowitz and I.A. Stegun (1964) *Handbook of Mathematical Functions* (Dover, New York).

Chapter 4

Entanglement, Nonlocality, Superluminal Signaling and Cloning

GianCarlo Ghirardi

Additional information is available at the end of the chapter

1. Introduction

Entanglement has been considered by E. Schrödinger [1] as: *The most characteristic trait of Quantum Mechanics, the one which enforces its entire departure from classical lines of thought.* Actually, the just mentioned unavoidable departure from the classical worldview raises some serious problems when entanglement of far away quantum systems is considered in conjunction with the measurement process on one of the constituents. These worries have been, once more, expressed with great lucidity by Schrödinger himself [1]: *It is rather discomforting that the theory should allow a system to be steered or piloted into one or the other type of state at the experimenter's mercy, in spite of his having no access to it.*

All those who are familiar with quantum theory will have perfectly clear the formal and physical aspects to which the above sentences make clear reference: they consist in the fact that, when dealing with a composite quantum system whose constituents are entangled and far apart, the free will choice of an observer to perform a measurement at one wing of the apparatus and the quantum reduction postulate imply that the far away state "jumps" in a state which depends crucially from the free will choice of the observer performing the measurement and from the random outcome he has got. Just to present a quite elementary case, let us consider a quantum composite system $S = S^{(1)} + S^{(2)}$, in an entangled state $|\psi(1,2)\rangle$:

$$|\psi(1,2)\rangle = \sum_i p_i |\phi_i^{(1)}\rangle \otimes |\gamma_i^{(2)}\rangle, \quad p_i \geq 0, \quad \sum_i p_i = 1. \tag{1}$$

In this equation (the Schmidt biorthonormal decomposition) the sets $\{|\phi_i^{(1)}\rangle\}$ and $\{|\gamma_i^{(2)}\rangle\}$ are two orthonormal sets of the Hilbert spaces of system $S^{(1)}$ and $S^{(2)}$, respectively, and, as such, they are eigenstates of appropriate observables $\Phi^{(1)}$ and $\Gamma^{(2)}$ of such subsystems. Suppose now that subsystem $S^{(2)}$ is subjected to a measurement of the observable $\Gamma^{(2)}$ and suppose that in the measurement the outcome (one of its eigenvalues) $\Gamma^{(2)} = g_r$ is obtained. Then, reduction of the wave packet leads instantly to the state $|\phi_r^{(1)}\rangle \otimes |\gamma_r^{(2)}\rangle$ for which one can claim that if system $S^{(1)}$ is subjected to a measurement of the observable $\Phi^{(1)}$ the outcome $\Phi^{(1)} = f_r$ will occur with certainty. Since this outcome, before the measurement on system $S^{(2)}$, has a nonepistemic probability p_r^2 of occurrence, one can state that the observation of $\Gamma^{(2)}$ has caused the instantaneous emergence at-a-distance of a definite property (which, according to quantum mechanics, one cannot consider as possessed in advance) of subsystem $S^{(1)}$, i.e. the one associated to the eigenvalue f_r of the observable $\Phi^{(1)}$. An analogous argument can obviously be developed without making reference to the Schmidt decomposition but to an arbitrary measurement on subsystem $S^{(2)}$, and will lead, in general, to the emergence of a different property of subsystem $S^{(1)}$ (typically the outcome of another appropriate measurement on this system becomes certain).

The just described process makes perfectly clear the nonlocal character of quantum mechanics, a fact that subsequently has been precisely identified by the illuminating work of J.S. Bell [2].

The situation we have just described will allow the reader to understand how it has given rise to the so called problem of faster-than-light signaling. If my action on system B, which takes place and is completed at a space like separation from system A, affects this system making instantaneously actual one of its potentialities, I can hope to be able to take advantage of this quantum peculiarity to make the observer A aware of the fact that I am performing some precise action on subsystem B at a space-like separation from him.

And, actually, this is what happened. From the seventies up to now, an innumerable set of proposals of taking advantage of entanglement and the reduction of the wave packet to achieve superluminal communication between distant observers appeared in the literature, proposals aiming to exploit this exciting possibility and to put into evidence the incompatibility of quantum mechanics with special relativity. Fortunately, as I will show in this paper, all proposal advanced so far, and, in view of some general theorems I will discuss below, all conceivable proposals of this kind, can been proven to be basically flawed in a way or another.

This chapter is devoted to discuss this important and historically crucial aspect of modern physics. As such, it has more an historical than a research interest. However, I believe that the reconsideration of the debate about this issue will be useful for the reader since many not so well known and subtle aspects of quantum mechanics will enter into play.

Before coming to a sketchy outline of the organization of the whole paper I would like to call the attention of the reader to a quite peculiar fact. When the so-called quantum measurement problem arose and was formalized by J. von Neumann [3], the attention of the scientific community was not concentrated on the possible conflicts between quantum mechanics and relativity; quantum mechanics was considered as a fundamentally nonrelativistic description of natural processes. Obviously, everybody had clear that the problem of its relativistic generalization had to be faced, but the debate concerned the nonrelativistic aspects of the theory and nobody had raised the question of possible conflicts between the two pillars of modern science[1]. In spite of that, the cleverly devised prescription of wave packet reduction, which was elaborated without having in mind relativistic potential oddities, turned out to be such that, in spite of its nonlocal nature and instantaneity, it did not allow to violate the basic relativistic request of no-faster-than-light signaling.

A brief outline of the organization of the chapter follows. After recalling the relevant aspects of the way in which quantum mechanics accounts for natural processes, we will describe various proposals for achieving faster-than-light signaling which have been put forward, and point out the reasons for which they are basically incorrect. To conclude this part we will present a general theorem showing that quantum mechanics, in its standard version, cannot in principle lead to superluminal communication.

However, the most interesting part of the debate is not the one we have just mentioned. In 1982 an analogous but quite different proposal of faster-than-light signaling has been put forward by N. Herbert [4]. The idea consisted in taking, as usual, advantage of the entanglement of far away subsystems and of wave packet reduction, but a new device was called into play: an hypothetical machine which could perform the task of creating many copies of an arbitrary state of a quantum system (a sort of "quantum xeroxing machine"). The interesting point is that at that time no general argument had been developed proving this task impossible. So, the mistaken suggestion of Herbert triggered the derivation of a theorem, the so-called no-cloning theorem, which was not known and which represents a quite relevant achievement which stays at the very basis of many important recent developments and which, besides proving that Herbert's proposal was unviable, plays a fundamental role for quantum cryptography and quantum computation.

As it is obvious, an hypothetical quantum device allowing faster-than-light communication would give rise to a direct and serious conflict with the special theory of relativity. As already stated, such a device is excluded by quantum mechanics. This, however, does not eliminate completely the potential tension of the nonlocal nature of quantum theory with the basic principles of relativity theory. The central issue is that the instantaneous collapse of the statevector of the far away system, even if it cannot be used to transfer energy or information at a superluminal speed, indicates that, in a way or another, an action performed in a given space-time region has some "effect" on systems at a space-like separation. Einstein has qualified this aspect of the theory as "a spooky action at-a-distance" which he could not accept. A. Shimony, by stressing the fact that the theory cannot be used to actually communicate superluminally has expressed his opinion that there is some sort of "peaceful coexistence of quantum mechanics and relativity" and has suggested to speak, in place of an "action" of a "passion" at-a-distance, to stress the peculiar nature of the perfect correlations of the outcomes, which, before any measurement process, individually have a fundamentally random nature, i.e., only a certain probability of occurrence.

Recently, the just mentioned problem has seen a revival due to the elaboration of the so-called "collapse models", i.e., modifications of quantum mechanics which, on the basis of a unique, universal dynamical principle account both for the quantum evolution of microscopic

---

[1] For this one should wait the celebrated EPR paper, which appeared, just as Schödinger's paper [1] in which the instantaneity of the reduction is seen as problematic, 2 years after von Neumann's precise formalization of the effect of a measurement.

systems as well as for the reduction process when macroscopic systems enter into play. Such theories, the best known of which is the one presented in ref.[5] usually quoted as "The GRW Theory", have been worked out with the aim of solving the macro-objectification or measurement problem at the nonrelativistic level, and the fact that they get the desired result in a clean, mathematically rigorous and conceptually precise way has raised the interest of various scientists, among them the one of Bell [6-8]. After the complete formalization of such approaches, it has been natural to start investigating whether they admit relativistic generalizations. Since they, agreeing with the quantum predictions concerning microsystems, exhibit (essentially) the same nonlocal aspects as standard quantum mechanics, the question of wether they actually can be made compatible with relativity has attracted a lot of attention. The serious work of various physicists in recent years has made clear that the program can be pursued, which means that one can have a theory inducing instantaneous collapses at-a-distance which does not violate any relativistic request. We consider it interesting to devote the conclusive part of this chapter to outline the investigations along these lines and to discuss their compatibility with the principles of special relativity.

2. The relevant formal aspects of the theory

2.1. The general rules

As is well known, quantum mechanics asserts that the most accurate specification of the state of a physical system is given by the statevector $|\Psi\rangle$, an element of the Hilbert space $\mathcal{H}$ associated to the system itself. When one deals with a statistical ensemble of identical systems, an equivalent and practical mathematical object is the statistical operator ρ which is the weighted sum of the statistical operators $|\psi_i\rangle\langle\psi_i|$ corresponding to the pure states $|\psi_i\rangle$ of the members of the ensemble: $\rho = \sum_i p_i|\psi_i\rangle\langle\psi_i|$, with $p_i \geq 0$, $\sum_i p_i = 1$. For an homogeneous ensemble or an individual system in a pure state $|\Psi\rangle$, the statistical operator is a projection operator: $\rho = |\Psi\rangle\langle\Psi|$.

The observables of the theory are represented by self-adjoint operators of the Hilbert space $\mathcal{H}$, which are characterized by their eigenvalues and eigenvectors. For the observable Ω one writes its eigenvalue equation as [2]:

$$\Omega|\omega_{k,\alpha}\rangle = \omega_k|\omega_{k,\alpha}\rangle, \tag{2}$$

where the index α is associated to the possible degeneracy of the eigenvalue ω_k. A crucial feature implied by the assumption of self-adjointness of the operators representing physical observables is that their spectral family, i.e. the projection operators $P_r = \sum_\alpha |\omega_{r,\alpha}\rangle\langle\omega_{r,\alpha}|$ on their eigenmanifolds, correspond to a resolution of the identity: $\sum_r P_r = I$, I being the identity operator on $\mathcal{H}$.

For what concerns the physical predictions, it is stipulated that in the case of a system in a pure state $|\Psi\rangle$ one has to express it as a linear combination of the eigensates of the observable (let us call it Ω) corresponding to the microscopic physical quantity which one intends to

---

[2] For simplicity we will deal with observables with a purely discrete spectrum, the changes for the continuous case been obvious.

measure: $|\Psi\rangle = \sum_{k,\alpha} c_{k,\alpha} |\omega_{k,\alpha}\rangle$. Then the theory asserts that the probability $P(\Omega = \omega_r | \Psi)$ of getting the outcome ω_r in the measurement of Ω when the system is in the pure state $|\Psi\rangle$, is given by $\sum_{\alpha} |c_{r,\alpha}|^2$, a quantity which coincides with the square of the norm $||P_r|\Psi\rangle||^2$ of the projection of the state onto the relevant eigenmanifold . This rule becomes, in the statistical operator language, $P(\Omega = \omega_r | \rho) = Tr[P_r \rho]$, where the symbol Tr means that the sum of the diagonal elements of the quantity in square brackets in an arbitrary orthonormal complete basis of $\mathcal{H}$ must be taken (this sum is easily proved not to depend on the chosen basis). Note that, using the complete set of the eigenstates of an operator Ω to evaluate the Trace, one immediately sees that its quantum average can be simply expressed as $\langle\Omega\rangle = Tr[\Omega\rho]$. It is an important mathematical fact that the Trace operation is linear and enjoys of the following formal feature: given two arbitrary (bounded) operators Λ and Γ of $\mathcal{H}$, $Tr[\Lambda\Gamma] = Tr[\Gamma\Lambda]$.

Before concluding this subsection we must also mention the effect on the statevector of performing a measurement process. Actually, two kinds of measurements can be carried out: the nonselective and the selective ones, i.e. those in which one measures an observable without isolating the cases in which a precise eigenvalue is obtained or, alternatively, those in which one is interested only in a definite outcome. They are represented, in the statistical operator language, by the two following formal expressions:

$$\rho_{before} \to \rho_{after} = \sum_k P_k \rho_{before} P_k, \tag{3}$$

$$\rho_{before} \to \rho_{after} = P_k \rho_{before} P_k / Tr[P_k \rho_{before}], \tag{4}$$

with obvious meaning of the symbols.

2.2. Composite systems

From now on we will be mainly interested in dealing with quantum systems S composed of two constituents, $S^{(1)}$ and $S^{(2)}$. Accordingly, their statevector $|\Psi(1,2)\rangle$ is an element of the tensor product $\mathcal{H}^{(1)} \otimes \mathcal{H}^{(2)}$ of the Hibert spaces of the constituents. As is well known, in the considered case two radically different situations may occur: in the first one the statevector is simply the direct product of precise statevectors for the constituents $|\Psi(1,2)\rangle = |\phi(1)\rangle \otimes |\gamma(2)\rangle$, and in such a case both constituents possess precise physical properties; alternatively, the statevector is entangled, i.e., it cannot be written in this form but it involves the superposition of factorized states, typically $|\Psi(1,2)\rangle = \sum_i c_i |\phi_i(1)\rangle \otimes |\gamma_i(2)\rangle$.

An extremely important point concerning composite systems is the following. Suppose one has a composite system and he is interested only in the outcomes of perspective measurement processes on one of the constituents. Then, one can easily convince himself that the simplest way of dealing with this problem is to consider the reduced statistical operator $\tilde{\rho}(1)$, obtained by taking the partial trace of the full statistical operator on the Hilbert space of the subsystem $S^{(2)}$ one is not interested in. At this point, to evaluate the probability of the outcomes of measurements of observables of the system of interest $S^{(1)}$, one can use the reduced statistical operator and the same prescriptions we have used for the general case [3] :

$$\tilde{\rho}(1) = Tr^{(2)}[\rho(1,2)]; \quad P(\Phi^{(1)} = f_r | \rho(1,2)) = Tr^{(1+2)}[P_r^{(1)}\rho(1,2)] \equiv Tr^{(1)}[P_r^{(1)}\tilde{\rho}(1)];$$
$$\langle \Phi^{(1)} \rangle = Tr^{(1)}[\Phi^{(1)}\tilde{\rho}(1)]. \tag{5}$$

It goes without saying that the operator $P_r^{(1)}$ in the previous equation is the projection operator onto the linear eigenmanifold associated to the eigenvalue f_r of $\Phi^{(1)}$.

2.3. von Neumann's ideal measurement scheme and its limitations

For the subsequent analysis it is important to briefly recall the so-called Ideal Measurement Scheme introduced by von Neumann in his celebrated book, ref.[3], and its limitations. The idea is quite simple: we are interested in "measuring" a microscopic observable, which is not directly accessible to our senses. Suppose then we have a microsystem s in a state $|\varphi_i^{(s)}\rangle$ which is in an eigenstate of a micro-observable $\Sigma^{(s)}$ pertaining to the eigenvalue s_i. How can one ascertain such a value, which, if a measurement is performed, according to the quantum rules will be obtained with certainty ? Von Neumann assumed that there exists a macroscopic object M which can be prepared in a ready state $|m_0\rangle$ and can be put into interaction with the microsystem. The interaction leaves unaltered the microstate while it induces, in a quite short time interval, the following evolution of the microsystem+apparatus:

$$|\varphi_i\rangle \otimes |m_0\rangle \rightarrow |\varphi_i\rangle \otimes |m_i\rangle, \tag{6}$$

where the states $|m_i\rangle$ are assumed to be orthogonal ($\langle m_i | m_j \rangle = \delta_{i,j}$), macroscopically and perceptively different (typically they are associated to different locations of the pointer of the macro-apparatus). Then, an observer, by looking at the measuring apparatus gets immediately the desired information concerning the value (s_i) of the microvariable.

The scheme is usually qualified as ideal because, in practice, the final apparatus states are not perfectly orthogonal and because very often the state of the microsystem is disturbed (or even the system is absorbed) in the measurement. The just mentioned scheme has an immediate important implication; the validity of Eq.(6) and the linear nature of Schrödinger's evolution equation imply that if one triggers the macroapparatus in its ready state with a superposition of the eigenstates of $\Sigma^{(s)}$, one has:

$$\sum_i c_i |\varphi_i\rangle \otimes |m_0\rangle \rightarrow \sum_i c_i |\varphi_i\rangle \otimes |m_i\rangle, \tag{7}$$

which is an entangled state of the microsystem and the macroapparatus.

---

[3] An elementary way to see this is to evaluate the probabilities of the joint outcomes of the measurement of a pair of observables, one for system $S^{(1)}$ and one for $S^{(2)}$, and then to sum on all possible outcomes of the measurement on the system we are not interested in.

Eq.(7) has given rise to one of the most debated problems of quantum mechanics, the so-called measurement or macro-objectification problem. In fact its r.h.s. corresponds to an entangled state of the system and the apparatus and in no way whatsoever to a state corresponding to a precise outcome[4] . The orthodox way out from this puzzle consists in resorting to the postulate of wave packet reduction: when a superposition of different macrostates emerges, a sudden change of the statevector occurs, so that one has to replace the r.h.s. of the previous equation with one of its terms, let us say $|\varphi_j\rangle \otimes |m_j\rangle$. This specific reduction occurs with probability $|c_j|^2$. We will not enter, here, in this deep debate, we simply mention that it amounts to accept (as many scientists did) that the linear character of the theory is violated (the reduction process is nonlinear and stochastic while the quantum evolution is linear and deterministic) at an appropriate (but not precisely specified) macroscopic level. Incidentally, von Neumann himself has proposed that the transition from the superposition to one of its terms occurs when a conscious observer becomes aware of the outcome (reduction by consciousness). Recently, various proposals of theories which, on the basis of a unique dynamical principle, account both for the linear nature of the evolution at the microscopic level as well as for the discontinuous changes (collapses) occurring when macrosystems are involved, have been put forward. We refer the reader to ref.[10] for an exhaustive analysis of such model theories.

2.3.1. Limits to the ideal scheme due to additive conservation laws

There are limitations to the von Neumann ideal scheme that we must mention because they have played a role in the refutation of some proposals of faster-than-light communication. Such limitations have been identified by Wigner [11], Araki and Yanase [12,13] in a series of interesting papers and subsequently they have been generalized in refs.[14,15]. The analysis by these authors takes into account the existence of additive conserved quantities for the system+apparatus system to derive precise conditions on a process like the one of Eq.(6) which stays at the basis of the von Neumann treatment. Let us summarize the procedure in a sketchy way. The process described by Eq.(6) represents the unitary evolution of the system+apparatus during the measurement process of the observable $\Sigma^{(s)}$ with eigenstates $|\varphi_i\rangle$. Let us therefore write it as: $U|\varphi_i, m_0\rangle = |\varphi_i, m_i\rangle$. Let us suppose that there exists an additive conserved quantity $\Gamma = \gamma^{(s)} \otimes I^{(A)} + I^{(s)} \otimes \gamma^{(A)}$ of the whole system and let us evaluate the matrix element of Γ, $\langle \varphi_i, m_0|\Gamma|\varphi_j, m_0\rangle$ by taking into account that Γ commutes with U, which implies :

$$\langle \varphi_i, m_0|(\gamma^{(s)} \otimes I^{(A)} + I^{(s)} \otimes \gamma^{(A)})|\varphi_j m_0\rangle = \langle \varphi_i, m_0|U^\dagger(\gamma^{(s)} \otimes I^{(A)} + I^{(s)} \otimes \gamma^{(A)})U|\varphi_j m_0\rangle. \quad (8)$$

---

[4] Note that in ref.[9] it has been proven that the occurrence of the embarrassing superpositions of macroscopically different states does not require that the measurement proceeds according to the ideal scheme of von Neumann. The same conclusion can be derived as a consequence of the necessary request that quantum mechanics governes the whole process and that one can perform a reasonably reliable measurement ascertaining the microproperty of the measured system.

We then have:

$$\langle \varphi_i, m_0 | (\gamma^{(s)} \otimes I^{(A)} + I^{(s)} \otimes \gamma^{(A)}) | \varphi_j m_0 \rangle = \langle \varphi_i | \gamma^{(s)} | \varphi_j \rangle + \delta_{ij} \langle m_0 | \gamma^{(A)} | m_0 \rangle =$$
$$\langle \varphi_i, m_0 | U^\dagger (\gamma^{(s)} \otimes I^{(A)} + I^{(s)} \otimes \gamma^{(A)}) U | \varphi_j m_0 \rangle = \langle \varphi_i, m_i | (\gamma^{(s)} \otimes I^{(A)} + I^{(s)} \otimes \gamma^{(A)}) | \varphi_j, m_j \rangle =$$
$$\delta_{ij} \langle \varphi_i | \gamma^{(s)} | \varphi_j \rangle + \delta_{ij} \langle m_i | \gamma^{(A)} | m_j \rangle. \tag{9}$$

Comparison of the final expression with the one after the equality sign in the first line shows that, in the considered case, one must have, for $i \neq j$, $\langle \varphi_i | \gamma^{(s)} | \varphi_j \rangle = 0$ which amounts to the condition that the observable $\Sigma^{(s)}$ which we want to measure on the microsystem must commute with the microsystem part $\gamma^{(s)}$ of the conserved additive quantity. If this is not the case (as it happens when $\Sigma^{(s)}$ is a component of the angular momentum of the system which does not commute with the other components), a process like the one of Eq.(6) turns out to be impossible; terms must be added to the r.h.s. involving other states of the microsystem besides $|\varphi_i\rangle$ and also other states of the apparatus. In refs.[13-15] it has been shown that in order to go as near as possible to the ideal case one must make more and more large the square of the norm of the state $\gamma^{(A)} | m_0 \rangle$. In the case of an angular momentum measurement this means to make the mean value of the square of the angular momentum component extremely large. Actually, in the case of the measurement of the spin component of a spin 1/2 particle, the "distorsion" of the state by the measurement, a quantity which can be estimated by the squared norm ϵ^2 of the state which has to be added to the r.h.s. of Eq.(6), must satisfy: $\epsilon^2 \geq h^2 / 32\pi^2 \langle m_0 | L^2 | m_0 \rangle$, where L^2 is the square of the angular momentum operator of the apparatus: to make the error extremely small one has to make extremely large $\langle m_0 | L^2 | m_0 \rangle$.

2.4. More realistic formalizations of the measurement process

Up to this point, when accounting for the occurrence of measurement processes, we have always made reference to the projection operators on the eigenmanifolds of the operators associated to the measurement. However, in practice, it is quite difficult to have apparatuses whose effect on the statevector can be accounted precisely by a projection operator. A simple example is the one of a detector of the position of a particle in a given interval Δ which has different efficiency in different portions of the interval Δ so that it detects for sure a particle impinging on its central region but only with a certain probability a particle which is detected near its extreme points. Another example is given by a measurement process which corresponds to two different successive measurements of two noncommuting observables, the outcome being represented by the pair of results which have been obtained. Also in this case the probability of "an outcome" cannot be expressed in terms of a single projection operator. The appropriate consideration of situations like those just mentioned has led to the consideration of more general processes affecting the statistical operator than the one of Eq.(3). One can then take advantage of a fundamental theorem by Kraus [16] asserting that the most general map of trace class and trace one semipositive definite operators onto themselves which respects also the condition of complete positivity (which has strong physical reasons to be imposed[5]) has the form:

---

[5] For a definition and a discussion of completely positive maps we refer the reader to ref.[16]

$$\rho \to \sum_i A_i^\dagger \rho A_i, \quad \sum_i A_i A_i^\dagger = I. \tag{10}$$

When considering measurement processes we will make reference to Eq.(3) or to the just written equation as expressing the effect of the measurement on the statistical operator.

3. Proposals of faster-than-light communication and their rebuttal

As already anticipated, after the clear cut proof by J.S. Bell of the fundamentally nonlocal nature of physical processes involving far away constituents in an entangled state, many proposals have been put forward, either in private correspondence or in scientific papers, suggesting how to put into evidence superluminal effects. We will begin by reviewing a series of proposal whose rebuttal did require only to resort to the standard formalism or to well established facts, such as those put into evidence by the Wigner-Araki-Yanase theorems.

3.1. Proposals taking advantage of the conservation of angular momentum

In the year 1979 various papers appeared asserting the possibility of superluminal communication by taking advantage of the change in the angular momentum of a far away constituent due to a measurement performed on its partner. The scientific and social context of these first investigations aiming to take advantage of quantum nonlocality have been described in the interesting and funny book [17] by D. Kaiser *How the hippies saved physics*, which intends to point out how the actions of a peculiar community of scientists and non scientists trying to justify various sort of paranormal effects on the basis of quantum nonlocality have drawn, in the US, the attention of the scientific community to Bell's fundamental theorem and its implications. The three papers [18-20] that I intend to consider in this section have some strict links with the just mentioned context.

Let us start with refs.[18,19]. Their argument is quite straightforward: one considers two far away spin 1/2 particles in the singlet state which interact with 2 apparatuses aimed to measure the spin z-component and are in their "ready" states $|A_0\rangle$ and $|B_0\rangle$, so that the initial state is:

$$|\Psi\rangle = \frac{1}{\sqrt{2}}[|1_+, 2_-\rangle - |1_-, 2_+\rangle] \otimes |A_0, B_0\rangle. \tag{11}$$

Here the indices + and - denote the values (in the usual units) of the z-component of the spin of the particles. Suppose now that the interaction of particle 2 with the apparatus B takes place before the other particle reaches A (A and B being at rest in a given inertial frame). Wave packet reduction occurs, and we are left, with the same probability, with one of the two states $|1_+, 2_-, A_0, B_-\rangle$ and $|1_-, 2_+, A_0, B_+\rangle$, where $|B_\pm\rangle$ are the states of the apparatus B after the measurement. We can now evaluate the mean value of the square of the spin angular momentum when the state is the one of Eq.(11) and when it is one of the states of the mixture. In the first case we get: $\langle S^2 \rangle_{singlet} = 0$, while in the second case we get the value $\hbar^2$. Now one takes advantage of the conservation of the angular momentum $\mathbf{L} = \mathbf{M} + \mathbf{S}$

where $\mathbf{M}$ is the angular momentum of the apparatus[6]. Since $\langle \mathbf{M} \cdot \mathbf{S} \rangle = 0$ in all above states, one concludes that the measurement induces a change of $\hbar^2$ in the apparatus which performs the measurement. This is not the whole story. If one, subsequently, leaves the second particle to interact with the apparatus measuring the spin state of the particle, the expectation value of $\mathbf{L}^2$ does not change any more. So, actually, the angular momentum of the apparatus which is the first to perform the measurement changes of the indicated amount, while the one of the other remains unchanged. Now if Alice and Bob, sitting near A and B, have at their disposal a source of entangled particles in the singlet state, Bob, who interacts first with his particle, can choose to perform or not the measurement; correspondingly he can choose whether to leave unchanged or to change the angular momentum of the apparatus at A. If Alice can detect this change she can get information about the choice made, in each single instance, by Bob[7]. Superluminal communication becomes possible.

According to the above analysis and the remark in the footnote, the key ingredient which allows to draw the conclusion is the occurrence of an ideal nondistorting measurement of the spin component. This implies that the argument of refs.[18,19] is based on contradictory assumptions, since, as discussed in the previous section, the Wigner-Araki-Yanase theorem asserts that the occurrence of an ideal nondistorting measurement of a spin component of a subsystem contradicts the conservation of total angular momentum. Actually, to have an ideal measurement process one needs apparatuses with a divergent mean value of the square of the angular momentum, but then no change of this quantity can be detected. Alternatively, one should consider nonideal measurements which are compatible with angular momentum conservation, but then the previous argument does not work, just because Eq.(6) has to be modified.

Precisely in the same year in which the above described arguments were presented, N. Herbert circulated a paper [20] which made resort to the functioning of a half wave plate to get the same result. His proposal was stimulated by his reading of a paper [21] written in 1936 by R. Beth and included by the American Association of Physics Teachers in a collection of papers published as *Quantum and Statistical Aspects of Light*. Beth managed to measure the angular momentum of circularly polarized light due to the fact that when right-circularly polarized light is shone on the half-wave plate it sets the plate spinning in one direction, while left-circularly polarized light spun the half-wave plate in the opposite direction. Moreover, the plate flips the light polarization from left to right and viceversa. Beth had measured the effect for circularly polarized light waves, i.e., by using a huge collection of photons all acting together. Herbert, inspired by this work, suggested, in the paper he called QUICK, to play a similar game with the angular momentum of individual photons to get superluminal effects.

Once more the idea is quite simple: one imagines a source emitting pairs of entangled photons in two opposite directions, their state being the analogous of the singlet state, i.e. the rotationally invariant state: $|\Psi(1,2)\rangle = [1/\sqrt{2}][|H1, H2\rangle + |V1, V2\rangle] \equiv [1/\sqrt{2}][|R1, L2\rangle + |L1, R2\rangle]$. Here, the symbols H, V, R and L make reference to the states of horizontal, vertical, right and left circular polarizations, respectively. Bob can freely choose whether to perform

---

[6] Here, we disregard the orbital angular momentum of the particles, but the argument holds true also without this limitation.

[7] It is interesting to remark that the same argument can be developed if one does not take into account the reduction process, i.e., if one assumes that the interactions simply take place in accordance with the von Neumann scheme.

a measurement of either plane (H,V) or circular (L,R) polarization. As a consequence of his measurement the far away photon is projected either onto a state of plane or of circular polarization. Subsequently, this photon impinges on a half-wave plate (near Alice). Since the photon when plane polarized crosses the plate without transmitting any angular momentum to it, while, when circularly polarized, it imparts a change of $\pm 2\hbar$ to the angular momentum of the plate, if Alice is able to check whether his plate has not changed or has changed its angular momentum she can know what kind of measurement (H,V) or (L,R) Bob has chosen to perform in any single case. Once more, entanglement and reduction of the wave packet allow superluminal transmission of information.

To prove why also this suggestion is inviable one has to analyze the functioning of the half-wave plate. The nice fact is that, as proved in ref.[22], one can develope precisely an argument analogous to the one of Wigner-Araki-Yanase, to prove that a half-wave plate can work as indicated only if a violation of the angular momentum conservation occurs. But such a conservation is necessary for the argument, so, once more, the proposal is contradictory. No superluminal communication is possible by resorting to the QUICK device.

Herbert and the Fundamental Fysiks Group made all they could do to spread out Herbert's conclusions. The debate involved scientists like H. Stapp and P. Eberhard. In june 1979, Stapp challenged the idea, building on Eberhard's argument that statistical averages would wash out any non local effect. But Herbert had worked out his reasoning for individual photons, and the above objection turned out to be not relevant for settling the issue. In the same year we (I' Weber and myself), became acquainted with Herbert's, as well as with Selleri's and others proposals. Accordingly, we wrote the paper [22] which presents the conclusions I have just described concerning refs.[18-20]. Beth's important experiment worked just because the experimenter sent an enormous number of photons at the half-wave plate. But, at the single-photon level, to get the same result, the half-wave plate would have to be infinitely massive, and, as such, it could not be put into rotation by the passage of an individual photon. This conclusion can be made rigorous with a little of mathematics, as we did in ref.[22].

3.2. Popper enters the game

Mention should also be made of the position of K. Popper concerning the problem of faster-than-light communication. In some previous writings, but specifically in his famous book [23] *Quantum Theory and the Schism in Physics* he raised the question of the conflict between quantum theory and special relativity theory, due to his alleged claim that "if quantum mechanical predictions are correct", then one would be able to send superluminal signals putting into evidence a conflict between the two pillars of our conception of the world. Unfortunately he was (mistankingly) convinced that the quantum rules would imply an effect that they actually exclude (a fact which he missed completely to understand), and, consequently, in his opinion they would allow superluminal signaling in an appropriate experimental situation.

The idea is quite simple (see Fig1, a,b): we have two perfectly correlated (in position) particles propagating towards two arrays of detectors placed at left (L) and right (R) of the emitting source at almost equal distances from it. Two slits, orthogonal to the direction x in the figure, are placed at both sides, along the y-axis, before the array of the counters, and, initially, only the counters lying behind the opening of the slits get activated. Subsequently, the slit at R is narrowed so as to produce an uncertainty principle scatter of the momentum p_y, which appreciably increases the set of counters which are activated with a non-negligible probability (see Fig.1b). Popper then argues: If quantum mechanics is correct, any increase in the knowledge of the position y at R like the one we get by making more precise the location along the y-axis of the particle which is there, implies an analogous increase of the knowledge of the position of the particle at L. As a consequence also the scatter at L should increase even though the width of the slit at this side has not been narrowed. This prediction is testable, since new counters would be activated with an appreciable probability, giving rise to a superluminal influence: Alice can know (with an appreciable probability) whether Bob has chosen to narrow or leave unchanged his slit. The conclusion of Popper is quite emblematic: in his opinion the increase of the spread at L would not occur and this *would show that quantum theory is wrong*. He also contemplates the other alternative: if the scatter at left would increase, then superluminal communication would be possible and relativity theory would be proven false; in both cases, a quite astonishing conclusion.

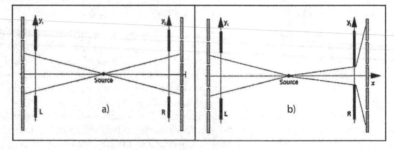

Figure 1. The set up and functioning of Popper's ideal experiment.

Unfortunately, in this passage of his important work, Popper shows his lack of understanding the quantum principles governing the unfolding of the considered experiment. In fact, it can be easily proved that quantum mechanics predicts precisely that no scatter at left will be induced by the narrowing of the slit at right. We do not consider it useful to enter in all technical details of the argument. The reader can look at ref.[24] or to Chapter 11 of ref.[25] for a detailed and punctual discussion. Here, we simply outline the argument: if the positions of the particles are really 100% correlated (and therefore associated to a Dirac delta like unnormalized state), then they are in a state which implies that, even when the two slits are fully opened, all counters are activated with large probabilities, while, if their correlations are only approximate (even with an extremely high degree of accuracy) the action at R by Bob does not change in any way whatsoever the outcomes at L. So, the argument is basically wrong.

4. The general proof of the impossibility of faster-than-light communication

To present a completely general proof [26,27] of the fact that instantaneous wave-packet reduction does not allow superluminal signaling we must start by reconsidering all possible actions [3,16] which are permitted, by standard quantum mechanics, on a constituent of a composite system. Quite in general, quantum mechanics allows the possibility of:

- A unitary transformation describing the free evolution of the system at R under consideration and/or possibly of its interactions with other systems lying in a space-time region which is space like with respect to the one at L.

- A transformation corresponding to a non selective projective measurement (with wave packet reduction as described by Eq. (3)) of the considered subsystem.

- A transformation like the one summarized in Eq.(10) corresponding, essentially, to the occurrence of a non ideal measurement.

- A transformation like the one of Eq.(4), corresponding to a selective measurement. To be strict, one should also consider the analogous of this transformation in the case of a non-ideal measurement, but, for the reasons which we will make clear below, this case does not have a physical relevance for faster-than-light signaling.

Now we can proceed to outline our proof, which represents a generalization and a more accurate formulation of some ideas put forward by P. Eberhard [28] about one year before we wrote our paper [26] (see also A. Shimony [29] for an enlightening discussion).

To start with we recall that all probabilistic predictions concerning a subsystem of a composite system can be obtained by considering the reduced statistical operator of the subsystem of interest. We suppose now to have a composite quantum system $S = S^{(1)} + S^{(2)}$ associated to the statistical operator $\rho(1,2)$ and to be interested in predictions concerning prospective measurements on subsystem $S^{(1)}$. As already remarked, the physics of this subsystem is fully described by the reduced statistical operator $\tilde{\rho}^{(1)} = Tr^{(2)}\rho(1,2)$. We can now consider the following set of equations:

$$\tilde{\rho}^{(1)} = Tr^{(2)}\rho(1,2),$$
$$Tr^{(2)}[U^{(2)\dagger}\rho(1,2)U^{(2)}] = Tr^{(2)}[U^{(2)}U^{(2)\dagger}\rho(1,2)] = Tr^{(2)}\rho(1,2) = \tilde{\rho}^{(1)},$$
$$Tr^{(2))}[\sum_i P_i^{(2)}\rho(1,2)P_i^{(2)}] = Tr^{(2)}[\sum_i [P_i^{(2)}]^2\rho(1,2)] = Tr^{(2)}\rho(1,2) = \tilde{\rho}^{(1)}$$
$$Tr^{(2))}[\sum_i A_i^{(2)\dagger}\rho(1,2)A_i^{(2)}] = Tr^{(2)}[\sum_i A_i^{(2)}A_i^{(2)\dagger}\rho(1,2)] = Tr^{(2)}\rho(1,2) = \tilde{\rho}^{(1)}. \tag{12}$$

In these equations we have made use of the cyclic property of the trace over the Hilbert space $\mathcal{H}^{(2)}$ when operators of the same Hilbert space are involved, of the unitarity relation $U^{(2)}U^{(2)\dagger} = I^{(2)}$, of the propery $[P_i^{(2)}]^2 = P_i^{(2)}$ and of the fact that the sum of the operators $P_i^{(2)}$ in the case of a projective measurement, as well as of the operators $A_i^{(2)}A_i^{(2)\dagger}$ of the Kraus theorem [16], must equal the identity operator of the same space. We call the attention of the reader to the fact that in all considered cases, i.e., i). no action on $S^{(2)}$, ii). a unitary evolution of $S^{(2)}$, or iii). the fact that it is subjected to an operation corresponding to a (ideal or non-ideal) nonselective measurement process, the reduced statistical operator $\tilde{\rho}^{(1)}$ of subsystem $S^{(1)}$ does not change in any way whatsoever, and, accordingly, Alice, performing measurements on such a subsystem cannot get any information about the fact that Bob is making some specific action on subsystem $S^{(2)}$.

Up to now, we have not considered explicitly the case of selective ideal or non-ideal measurement processes, accounted for by Eq. (4) or by its analogue referring to processes like those governed by Eq.(10). If one considers the modifications to the general statistical operator in these cases and one uses the reduced statistical operator to evaluate the probabilities of the measurement outcomes on subsystem $S^{(1)}$, one would easily discover that the physics of such a system is actually changed by the action on its far-away partner. But the probabilistic changes depend crucially on the outcome that Bob has obtained in his measurement, so that Alice might take advantage of this fact only if she would be informed of the outcome obtained by Bob. This implies that Bob must inform Alice concerning his outcome and this can be done only by resorting to standard communication procedures which require a subluminal communication between the two. Accordingly, these cases can safely be disregarded within our context.

It should be clear that the general validity of our theorem implies that all previously discussed attempts to get faster-than-light signaling taking advantage of the instantaneous reduction at-a-distance of the statevector in the case of entangled states of far-away systems, were doomed to fail. We have discussed them in some detail to present an historically complete perspective of the debate on this fundamental issue, i.e. the one of the compatibility of quantum mechanics with relativistic requirements concerning the communication between far-away observers.

5. A radical change of perspective

5.1. Herbert's new proposal

In 1981 N. Herbert submitted for publication to Foundations of Physics a paper [4] by the title: *FLASH–A superluminal communicator based on a new kind of quantum measurement* in which he added a new specific device to his previous proposal we have discussed in sect.3.1. The stimulus to do so came probably from our paper, ref.[26], as remarked by D. Kaiser in his book [17]: *From Ghirardi's intervention, Herbert came to appreciate the importance of amplifying the tiny distinction between various quantum states, to evade fundamental limits on signaling.* The crucial device which, in his opinion, could do the game, was a Laser gain tube exhibiting the following characteristics: if the laser was stimulated by a single photon in any state of polarization (the states which mattered were actually those of plane (V and H) and of circular (R and L) polarization) it would emit a relevant number, let us say 4N with N large, of identical copies (in particular with the same polarization) of the impinging photon. If we summarize the process by means of an arrow leading from the initial photon state to the bunch of final photons, Hebert's Laser gain tube has to work in the following way:

$$|V,1\rangle \to |V,4N\rangle, \ |H,1\rangle \to |H,4N\rangle, \ |R,1\rangle \to |R,4N\rangle, \ |L,1\rangle \to |L,4N\rangle, \quad (13)$$

with obvious meaning of the symbols. Here 1 denotes the photon propagating towards Alice.

By resorting to this machine Herbert's game becomes quite simple. One starts, as in his first proposal, with a source emitting pairs of entangled photons in two opposite directions, their state being the rotationally invariant state: $|\Psi(1,2)\rangle = [1/\sqrt{2}][|H1,H2\rangle + |V1,V2\rangle] \equiv [1/\sqrt{2}][|R1,L2\rangle + |L1,R2\rangle]$. Obviously, Bob can freely choose whether to perform a measurement of either plane (H,V) or circular (L,R) polarization. As a consequence of his measurement the far away photon is projected either onto a state of plane or of circular polarization. At this point the far away photon is injected in the Laser gain tube which emits 4N photons with the same polarization, which, in turn, depends on the measurement performed by Bob and the outcome he has got. The 4N photons are then separated into 4 beams of N photons each, directed towards 4 detectors of V,H,L and R (mind the order) polarization, respectively. To see the game coming at an end we have now simply to recall that a detector registers for sure a photon with the polarization it is devised to measure, it does not detect a photon into an orthogonal state and detects with probability $1/2$ a photon in a state of polarization which is the equal superposition of the state that it is devised to detect and of the state orthogonal to it.

We analyze the situation in detail specifying the measurements which Bob chooses to perform, the outcomes he gets and the records by the counters near Alice.

- Suppose Bob chooses to perform a polarization measurement aimed to ascertain whether the photon (2) reaching him has vertical or horizontal polarization and that he finds the photon vertically polarized. In this case, the process goes as follows:

  a). Initial state: $|\Psi(1,2)\rangle = [1/\sqrt{2}][|H1,H2\rangle + |V1,V2\rangle]$; b). Measurement with outcome Vertical; c). Reduction of the state: $|V1,V2\rangle$; d). Amplification: $|V,4N;V2\rangle$; e). Number of photons detected by the far away detectors (near Alice) for the 4 beams: N,0,N/2,N/?

- a). Initial state: $|\Psi(1,2)\rangle = [1/\sqrt{2}][|H1,H2\rangle + |V1,V2\rangle]$; b). Measurement with outcome Horizontal; c). Reduction of the state: $|H1,H2\rangle$; d). Amplification: $|H,4N;H2\rangle$; e). Number of photons detected by the far away detectors on the 4 beams: 0,N,N/2,N/2.

- a). Initial state: $[1/\sqrt{2}][|R1,L2\rangle + |L1,R2\rangle]$; b). Measurement with outcome Right; c). Reduction of the state: $|L1,R2\rangle$; d). Amplification: $|L,4N;R2\rangle$; e). Number of photons detected by the far away detectors on the 4 beams: N/2,N/2,N,0.

- a). Initial state: $[1/\sqrt{2}][|R1,L2\rangle + |L1,R2\rangle]$; b). Measurement with outcome Left; c). Reduction of the state: $|R1;L2\rangle$; d). Amplification: $|R,4N;L2\rangle$; e). Number of photons detected by the far away detectors on the 4 beams: N/2,N/2,0,N.

Now, one has simply to remark that in the cases listed under the two first items (i.e. when Bob chooses to measure linear polarization) the detector which does not register any photon is either the first or the second, while, in the alternative case in which Bob chooses to measure circular polarization, it is either the third or the fourth detector which does not register any photon. Accordingly, Alice can become aware, instantaneously, of the choice made by Bob: they can communicate superluminally.

5.2. The no-cloning theorem

The FLASH paper was sent for refereeing to A. Peres and to me. Peres' answer [30] was rather peculiar: *I recommended to the editor that this paper should be published. I wrote that it was obviously wrong, but I expected that it would elicit considerable interest and that finding the error would lead to significant progress in our understanding of physics.* I also was rather worried for various reasons. I was not an expert on Lasers and I was informed that A. Gozzini and R. Peierls were trying to disprove Herbert's conclusion by invoking the unavoidable noise affecting the Laser which would inhibit its desired functioning. On the other hand, I was convinced that quantum theory in its general formulation and not due to limitations of practical nature would make unviable Herbert's proposal. After worrying for some days about this problem I got the general answer: while it is possible to devise an ideal apparatus which clones two orthogonal states with 100% efficiency, the same apparatus, if the linear quantum theory governs its functioning, cannot clone states which are linear combinations of the previous ones. Here is my argument, on the basis of which I recommended rejection of Herbert's paper. The assumption that the cloning machine acts as follows:

$$|V\rangle \rightarrow |V, 4N\rangle \;\; and \;\; |H\rangle \rightarrow |H, 4N\rangle, \tag{14}$$

when the linear nature of the theory is taken into account, implies:

$$R \equiv \frac{1}{\sqrt{2}}[i|V\rangle + |H\rangle] \rightarrow \frac{1}{\sqrt{2}}[i|V, 4N\rangle + |H, 4N\rangle], \tag{15}$$

and analogously for the left polarization. Now, the state at the r.h.s. of the last equation is by no means the state $|R, 4N\rangle$ which Herbert had assumed to occur in the case in which the Laser gain tube is triggered by a right polarized photon. But this is not the whole story: how it has been shown in ref.[31] the very linear nature of the theory implies that no difference in the detections of Alice occurs in dependence of the free will choice of Bob.

This is an account of how Herbert's ingenious, but mistaken, proposal has led me to be the first to derive the no-cloning theorem [8]. About one and half year later Wootters and Zurek [32] and Dieks [33] derived independently the same result and published it[9]. The theorem is of remarkable importance in quantum theory, it has become known as "The no cloning theorem" and it has been quoted an innumerable number of times. Only subsequently I realized that it had been a mistake on my part not to publish my result. I discussed my precise argument with Gozzini and Peierls, by sending them a draft which was a sort of repetition of my referee report and I subsequently published it [31] in collaboration with my collaborator, T. Weber.

5.2.1. More on quantum cloning

In a paper like the present one, we believe it useful to mention that E.P. Wigner [34], in an essay of 1961 had already argued that the phenomena of self-replication of biological systems contradict the principles of quantum mechanics. His argument is quite straightforward. Following his notation let us suppose we have a living system in a state v and an environment (assimilated to "food") in a state w, so that, the initial statevector of the system, organism +

---

[8] I have chosen to attach at the end of the paper, a document - a letter by A. van der Merwe - which officially attests this fact, since it is known only to a restricted community of physicists.

[9] I must confess that I have never understood why A. Peres, in mentioning my derivation, has stated that it was a special case of the theorems in refs. [32] and [33]. Comparison even only of the short page by A. van der Merwe with the just mentioned papers makes clear that the argument is precisely the same and has the same generality.

nutrient, is: $\Phi = v \times w$. When replication takes place the statevector will have the form: $\Psi = v \times v \times r$, i.e. two organisms each in the statevector v will be present, while the vector r describes both the rest of the system, the rejected part of the nutrient and also the other coordinates (positions, etc.) of the two organisms. One assumes that the system lives in an N-dimensional Hilbert space $\mathcal{H}^{(\mathcal{N})}$, the part r in an R-dimensional Hilbert space $\mathcal{H}^{(\mathcal{R})}$, while the "food" state w belongs to a $N \cdot R$ dimensional Hilbert space $\mathcal{H}^{(\mathcal{NR})}$, so that Φ and Ψ live, as they must, in the same space. Suppose we do not know the state of the living system; however, since it belongs to $\mathcal{H}^{(\mathcal{N})}$ his knowledge requires to know N complex numbers. Analogously we do not know the state r, and the state w, which require the specification of R and NR complex numbers. We now assume, with Wigner, that the collision matrix which gives the final state resulting from the interaction, which will be denoted as S, of the organism and the nutrient is a random matrix, which, however, even though unknown to us explicitly, is completely determined by the laws of quantum mechanics. Obviously S must satisfy:

$$\Psi = S\Phi. \tag{16}$$

Choosing a basis for the whole Hilbert space and projecting Eq. (16) on such a basis one gets N^2R equations. And now the conclusion follows: our unknown quantities are the components of the states v, r and w on their respective bases and are therefore $N + R + NR$ in number. Thus, according to Wigner, the question is: given the matrix corresponding to S, it is possible to find vectors v, r and w such that their components satisfy the above mentioned N^2R equations? Since $N^2R \gg N + R + NR$, for extremely large N and R, according to him: *it would be a miracle if such equations could be satisfied.* In other words, a self-replicating quantum unit does not exist. One might state that Wigner has "derived" (with the proviso he is making - see below) the no-cloning theorem for a quantum system whose Hilbert space has an extremely high dimensionality N, while we have shown that it holds also for N=2.

Wigner was perfectly aware that the argument is not fully rigorous and cannot be taken too seriously because of the many assumptions on which it is based. However, he seems inclined to attach a certain value to it. This is not surprising because at the time in which he wrote his paper he was adhering to von Neumann's idea that consciousness is responsible for the reduction of the wave packet, so that, in a certain sense, the fact that quantum mechanics is not able to account for the basic property of living organism (the self reproduction) supported his view that such a theory cannot be used to describe the conscious perceptions of such organisms. In 1971 Eigen [36] responded to Wigner claiming that his choice of resorting to a typical unitary map to account for the process did not take into account the instructive functions of informational macromolecules.

Strictly connected with Wigner argument, even though derived through a much more rigorous and general procedure is the proof of the no-cloning theorem presented in a beautiful paper by R. Alicki [35]. He considers a dynamical transformation T from an initial state $\varphi \otimes \omega$, where φ is the state of the organism and ω the fixed initial state of the environment designed as "food":

$$\varphi \otimes \omega \rightarrow T(\varphi \otimes \omega) = \varphi \otimes \varphi \otimes \sigma. \tag{17}$$

As before, σ represents the state of the "food" after the replication. Alicki assumes that any dynamical process of a closed system (typically the one given by T) cannot reduce the indistinguishability of two states φ and ψ, which can be quantified by the "overlap" $(\varphi|\psi)$ of the two states[10], i.e., $(T(\varphi)|T(\psi)) \geq (\varphi|\psi)$ (which in Alicki's spirit can be considered as a form of the second law of thermodynamics: indistinguishability cannot decrease with the evolution). Then one has:

$$(\varphi|\varphi') = (\varphi \otimes \omega|\varphi' \otimes \omega) \leq (T(\varphi \otimes \omega)|T(\varphi' \otimes \omega))$$
$$= (\varphi \otimes \varphi \otimes \sigma|\varphi' \otimes \varphi' \otimes \sigma') = (\varphi|\varphi')^2(\sigma|\sigma') \leq (\varphi|\varphi')^2, \tag{18}$$

implying $(\varphi|\varphi') = 1$ or $(\varphi|\varphi') = 0$.

It is interesting to note that if, taking a strictly quantum perspective (which means to replace the round brackets in the above equation by Dirac's bras and kets), one identifies (as it is quite reasonable) the general concept of overlap≡indistinguishability with the scalar product of the Hilbert space and one assumes that the unfolding of the process is governed by a unitary transformation (which as such does not change the overlap), the above proof (slightly modified to take into account the complex nature of the scalar product) corresponds to a modern version of the no-cloning theorem which, in place of using the linearity of the evolution as we and the authors of refs.[32,33] did in deriving the theorem, makes resort to unitarity.

6. Further recent proposals which require new impossibility proofs

6.1. A proposal by D. Greenberger

In spite of the lively debate and the many precise results which should have made fully clear why quantum mechanics does not allow superluminal communication, new papers claiming to have found a new way to achieve this result continue to appear. The first we want to mention is a proposal [37] of D. Greenberger which has been considered as inspiring even quite recently. Actually, in ref.[38] it is claimed that the proposal of Greenberger *has not yet been refused and calls into question the universality of the no-signaling theorem*, and, accordingly, it represents a stimulus to pursue the investigations on this line.

Greenberger proposal involves the simultaneous emission of two photons by a source along two different opposite directions (a, a') and (b, b'), so that the initial state is the entangled state:

$$|\psi\rangle_{1,2} = \frac{1}{\sqrt{2}}[|a\rangle_1 \otimes |a'\rangle_2 + |b\rangle_1 \otimes |b'\rangle_2] \tag{19}$$

Subsequently, the two photons impinge on a series of beam splitters, as shown in figure 2. The horizontal gray boxes represent the beam splitters which are assumed to both reflect and transmit half the incident light, and produce a phase shift of $\pi/2$ upon reflection and none

---

[10] Here , the expressions $(\alpha|\beta)$ must not be identified with the Hilbert scalar product, since Alicki is taking a much more general perspective, which, however, requires to quantify the idea of distinguishability and its fundamental properties. He does this by introducing his symbol for the overlap which he takes, for simplicity, to be a real number between 0 and 1.

upon transmission. On the path of the photon emitted along b, after it goes through the first beam splitter, there is a phase-shifter A that shifts the phase of any photon passing through it by π, and that can be inserted or removed from the beam at will.

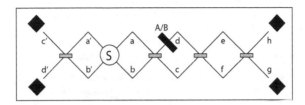

Figure 2. Illustration of Greenberger's proposal as depicted in his paper.

At this point the first crucial assumption of the paper enters into play:

i). *The phase shifter can be prepared not only in the states* $|A\rangle$ *and* $|B\rangle$, *corresponding to its being inserted or removed from the path of the photon, but also in their orthogonal linear combinations:*

$$|u\rangle_3 = \frac{1}{\sqrt{2}}[|A\rangle + |B\rangle], \qquad |v\rangle_3 = \frac{1}{\sqrt{2}}[|A\rangle - |B\rangle]. \tag{20}$$

According to the author of [37], one can also switch on the Hamiltonian H for this macroscopic object, whose eigenstates are $|u\rangle_3$ and $|v\rangle_3$, corresponding to slightly different energies, implying the development in time of relative phases with respect to each other.

We will not go through the subsequent elementary calculations of the paper; we limit ourselves to mention that the above assumptions lead to the conclusion that, as the photons are nearing their final detectors represented in the figure by the 4 black squares, they will be in the following entangled photon-phase shifter state:

$$|\psi\rangle_{1,2,3} = \frac{1}{2}[(-e^{i\alpha}|h\rangle_1|d'\rangle_2 + e^{-i\alpha}|g\rangle_1|c'\rangle_2)e^{i\beta}|u\rangle_3 + (e^{i\alpha}|g\rangle_1|c'\rangle_2 - e^{-i\alpha}|h\rangle_1|d'\rangle_2)e^{-i\beta}|v\rangle_3], \tag{21}$$

the phase factors $e^{\pm i\alpha}$ and $e^{\pm i\beta}$ being due to the evolution of the states $|u\rangle_3$ and $|v\rangle_3$.

At this point Greeneberger puts forward his really crucial assumption. In his words:

ii). *In accordance with our assumption that one can manipulate these Cat states, one can turn off H for the state* $|v\rangle$, *while leaving it in place for the state* $|u\rangle$. *This will rotate the state* $|v\rangle$ *into the state* $e^{i\gamma}|u\rangle$, *where* γ *is the accumulated phase difference during this process.*

As it is obvious this amounts to accept that a nonunitary transformation T can be performed:

$$T|u\rangle_3 = |u\rangle_3, \qquad T|v\rangle_3 = e^{i\gamma}|u\rangle_3. \tag{22}$$

The conclusion follows. After this transformation the state becomes:

$$|\psi_{final}\rangle_{1,2,3} = e^{i\gamma/2}\left[-\cos(\alpha+\beta-\gamma/2)|h\rangle_1|d'\rangle_2 + \cos(\beta-\alpha-\gamma/2)|g\rangle_1|c'\rangle_2\right]|u\rangle_3. \quad (23)$$

And now the game is over: by appropriately choosing the angles α, β and γ, one can, at his free will, suppress one of the two terms of the superposition of the photon states, i.e. one can make certain either the firing of the detector in d' or the one in c' (and correspondingly the one in h or the one in g) allowing in this way a superluminal transfer of information from the phase shifter, which acts as the signaler, to the photon detectors.

The paper, since nobody had discussed it in spite of its revolutionary character, deserved some attention; we have reconsidered it in ref.[39]. Its weak points are:

- The assumption that one can prepare the linear superposition of two macroscopically different states, corresponding to different locations of the macroscopic phase-shifter. This is impossible to get in practice.
- However, even ignoring the above critical feature of the hypothetical experiment, the really crucial and unacceptable fact is the one embodied in its second assumption, i.e., the possibility of implementing a nonunitary transformation.

We will not go on analyzing all the details of ref.[37] and of the punctual criticisms of ref.[39]. We believe that to show where it fails the simplest way is to resort to an example that we have devised in our paper. We consider an elementary EPR-Bohm like setup for two far away spin $1/2$ particles in the singlet state:

$$|\psi_-\rangle = \frac{1}{\sqrt{2}}[|\uparrow_1\rangle|\downarrow_2\rangle - |\downarrow_1\rangle|\uparrow_2\rangle]. \quad (24)$$

In strict analogy with what has been assumed by Greenberger, suppose now we can rotate only one of the two spin states of particle 2 making it to coincide, apart from a controllable phase, with the other one:

$$T|\downarrow_2\rangle = |\downarrow_2\rangle, \qquad T|\uparrow_2\rangle = e^{i\gamma}|\downarrow_2\rangle. \quad (25)$$

Under this transformation the state [25] becomes a factorized state of the two particles:

$$|\psi_T\rangle = \frac{1}{\sqrt{2}}[|\uparrow_1\rangle - e^{i\gamma}|\downarrow_1\rangle]|\downarrow_2\rangle \quad (26)$$

In (26), the state referring to particle 1 is an eigenstate of $\sigma \cdot \mathbf{d}$ for the direction $\mathbf{d} = (\cos\gamma, \sin\gamma, 0)$ pertaining to the eigenvalue -1. This means that a measurement of this observable by Alice (where particle 1 is) will give with certainty the outcome -1 if Bob has performed the transformation T on his particle, while, if Bob does nothing, the probability of getting such an outcome equals $1/2$. Having such a device, one can easily implement superluminal transfer of information. Concluding: if assumption ii) were correct, one would not need all the complex apparatus involved in Greenberger's proposal which, at any rate, cannot work as indicated due to the nonlinear nature of T.

6.2. A proposal involving a single system

Another proposal that has to be mentioned is the one [40] by Shiekh. His suggestion is different from all those which have appeared in the literature since the author does not make resort to an entangled state of two systems but he works with a single particle in a superposition of two states corresponding to its being in two far-away regions, and the measurement process involves only one of the two far-away parts of the wave function. So, in a sense, the argument of ref.[40] does not fall under the no-go theorems considered here and requires a separate comment. The author is inspired by the fact that when a single particle is associated to a wavefunction as the one just mentioned, any attempt to test whether it is "here" (at right), or "there" (at left) changes instantaneously the wavefunction on the whole real axis by making it equal to zero or enhancing it "there" according to whether we detect or we do not find the particle "here". The process seems to exhibit some nonlocal aspects due to the instantaneous change at-a-distance. Obviously, that this might lead to superluminal signaling is something that nobody can believe, but it is instructive to show that also in this case, to achieve the desired result, one has to resort to a nonunitary evolution. The elementary analysis which follows will lead once more to the conclusion that the process cannot be used to send superluminal signals.

We briefly review the argument by Shiekh. He considers a particle which is prepared, at time $t = 0$, in an equal weights superposition of two normalized states, $|h+\rangle$ and $|h-\rangle$, propagating in two opposite directions, respectively, starting from the common origin of the x-axis:

$$|\psi, 0\rangle = \frac{1}{\sqrt{2}}[|h+\rangle + |h-\rangle]. \tag{27}$$

Subsequently the state $|h+\rangle$ is injected in an appropriate device behaving in a way rather similar, apart from the final stage, to a Mach-Zender interferometer. One also assumes that an observer, located near to it, can choose, at his free will, to insert or not a phase-shifter along one of the two paths of the interferometer. The two wave functions are then recombined by appropriate deflectors so that, by deciding whether or not to insert the phase-shifter, one can produce a constructive (no phase-shifter in place) or a destructive (the phase-shifter is present) interference of the two terms in which the impinging state $|h+\rangle$ has been split. Finally, a detector is placed along the direction of propagation of the final state and it induces wave packet reduction, since it either detects or fails to detect the particle. We have summarized the situation for the two considered cases in Figs. 3a,b.

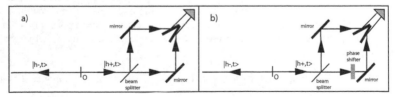

Figure 3. The experimental arrangement of ref.[40]. The two cases refer to no Phase-shifter inserted or phase shifter inserted, respectively.

The author then concludes: *If the sender* (the guy who can choose to insert or not the phase shifter), *arranges for constructive interference, then some of the particles will be "taken up" by the sender, but none if destructive interference is arranged; in this way the sender can control the intensity of the beam detected by the receiver* (the observer located far away where the evolved of $|h-\rangle$ is concentrated). *So, a faster than light transmitter of information (but not energy or matter) might be possible.*

We believe that all readers will have clear the trivial mistake of the paper. In fact, what one can govern by deciding whether to insert or not the phase-shifter, is the interference at the central region of the final detector. Let us concentrate our considerations only on the normalized state $|h+\rangle$. If triggered by such a state when it exhibits constructive interference, the counter will register (practically) for sure the particle described by such a state (the wavefunction has a peak just there), while, if there is destructive interference, the counter will not register the particle. But this does not mean that the wavefunction associated to $|h+\rangle$ disappears, as claimed by the author, it simply means that its support lies outside the interval covered by the counter. Actually, no one will doubt that if one places an array of counters covering all the line orthogonal to the final direction of propagation, one of them, different from the one of the experiment, will fire for sure. If one combines these considerations with the fact that actually the whole state of the particle is the superposition of $|h-\rangle$ and $|h+\rangle$ one realizes that the statements we have just made concerning what is going on at right have only probability $1/2$ of occurrence, since the particle can be not detected in the right region. Accordingly, the probability that the particle is found at left remains equal to $1/2$, as if no specific action would be made at right.

It seems rather peculiar that the author introduces an hypothetical process which can make zero a wavefunction (i.e. the normalized state $|h+\rangle$), and as such it does not preserve unitarity, and, at the same time, he resorts to the overall conservation of probability (i.e. to unitarity) to claim that the action at right can change the norm of the state at left.

This concludes our analysis of the many proposals which have been presented to send superluminal signals.

7. Nonlocality and relativistic requirements

As already stated, quantum mechanics suffers of an internal inconsistency, the one between the linear and deterministic evolution induced by Schrödinger's dynamics and the nonlinear and stochastic collapse of the state in a measurement. Many scientists, among them Einstein, Schrödinger, Bell and many others have been disturbed not only by the formal inconsistency between the two dynamical principles, but especially by the fact that the borderline between what is classical and what is quantum, what is reversible and what is irreversible, what is micro and what is macro is to a large extent ambiguous. Accordingly, many serious attempts to overcome this difficulty have been presented, inspired by the conviction that Bell has expressed [6] so lucidly:

> *Either the wavefunction, as given by the Schrödinger equation is not everything or it is not right.*

7.1. Bohmian mechanics

The first alternative corresponds to the idea that the specification of the state of a physical system given by the statevector has to be enriched or replaced by new variables (the so-called hidden variables). The best known and rigorous example of this line is represented by Bohmian mechanics [42], a deterministic theory such that the assignement of the wavefunction and of the hidden variables (i.e. the initial positions of all particles which are chosen to be distributed according to the quantum probability $|\psi(\mathbf{r}_1, \mathbf{r}_2, ..., \mathbf{r}_N, t_0)|^2$) determines uniquely their positions at any subsequent time. The predictions of the theory concerning the probability distribution of the particles coincide with those of standard quantum mechanics and the theory overcomes the measurement problem in a clean and logically consistent way.

I will spend only few words on the locality issue within Bohmian mechanics. Since this theory agrees with quantum mechanics in general and typically in an EPR-like situation, it must exhibit a specific sort of nonlocality. It has been proved [43] that any deterministic hidden variable theory equivalent to quantum mechanics admits only relativistic generalizations which must resort to a specific foliation of space-time. In other words, such theories are characterized by a preferred reference frame which, however, remains unaccessible. Accordingly, as stressed by Bell [8], they require a change of attitude concerning Lorentz invariance: the situation resembles the one of the theory of relativity in the Fitzgerald, Larmor, Lorentz and Poincaré formulation in which there is an absolute aether, but the contraction of space and the dilation of time fooled the moving observers by allowing them to consider themselves at rest. In spite of this remark, explicit and interesting relativistic generalizations of Bohmian mechanics have been presented. In particular bohmian-like relativistic models have been worked out both in first quantized versions [44] as well as in the framework of quantum field theories [45].

7.2. Collapse theories

The second alternative corresponds to assuming that Schrödinger's equation has to be changed in such a way not to alter the well established predictions of quantum mechanics for all microscopic systems while leading to the collapse of the statevector with the desired features and probabilities when macroscopic systems enter into play. The first explicit example of this kind is the so called GRW theory [5] which we summarize in a very sketchy way.

The central idea is to modify the linear and deterministic evolution equation of standard quantum mechanics by adding nonlinear and stochastic (i.e. sharing the features of the reduction process) terms to it. As it is obvious, and as it has been stressed by many scientists, since the situations characterizing macro-objects correspond to perceptually different locations of their macroscopic parts (e.g. the pointer) the change in the dynamics must strive to make definite the positions of macroscopic bodies. The model is based on the following assumptions:

- A Hilbert space $\mathcal{H}$ is associated to any physical system and the state of the system is represented by a normalized vector $|\psi_t\rangle$ of $\mathcal{H}$,

- The evolution of the system is governed by Schrödinger's equation. Moreover, at random times, with mean frequency[11] λ, each particle of any system is subjected to random

---

[11] Actually this frequency must be made proportional to the mass of the particles entering into play. The value we will choose below refers to nucleons.

spontaneous localization processes as follows. If particle i suffers a localization then the statevector changes according to:

$$|\psi_t\rangle \rightarrow \frac{L_i(\mathbf{x})|\psi_t\rangle}{\|\ L_i(\mathbf{x})|\psi_t\rangle\ \|}; \quad L_i(\mathbf{x}) = (\frac{\alpha}{\pi})^{3/4}e^{[-\frac{\alpha}{2}(\hat{x}_i-\mathbf{x})^2]}, \tag{28}$$

where $\hat{x}_i$ is the position operator of particle i,

- The probability density for a collapse at $\mathbf{x}$ is $p(\mathbf{x}) = \|\ L_i(\mathbf{x})|\psi_t\rangle\ \|^2$, so that localizations occur more frequently where the particle has a larger probability of being found in a standard position measurement.

The most relevant fact of the process is its "trigger mechanism", i.e. the fact that, as one can show by passing to the centre-of-mass and relative coordinates, the localization frequency of the c.o.m. of a composite system is amplified with the number of particles, while the internal motion, with the choice for α we will make, remains practically unaffected. We have summarized the situation for a micro (at left) and macroscopic (at right – a pointer) system in Fig.4.

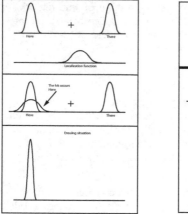 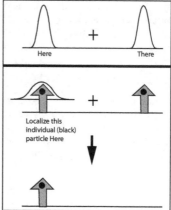

Figure 4. The localization of a single particle and of a macro-object according to GRW.

With these premises we can now make the choice for the parameters α (note that $\frac{1}{\sqrt{\alpha}}$ gives the localization accuracy) and λ of the theory. In ref.[5] we have chosen:

$$\alpha \simeq 10^{10}cm^{-2}, \lambda \simeq 10^{-16}sec^{-1}. \tag{29}$$

Note that with these choices a microscopic system suffers a localization about every 10^7 years, and this is why the theory agrees with quantum mechanics for such systems, a macroscopic body every $10^{-8}sec$ (due to the trigger mechanism in the case of an Avogadro number of particles the frequency becomes $10^{24} \cdot 10^{-16} = 10^8 sec^{-1}$). As commented [6] by J. Bell: *The cat is not both dead and alive for more than a split second.*

The conclusion should be obvious. A universal dynamics has been worked out which leaves (practically) unaltered all quantum predictions for microsystems but it accounts for wave packet reduction with probabilities in agreement with the quantum ones and for the classical behaviour of macroscopic systems, as well as for our definite perceptions concerning them.

Few remarks: i). The model has been generalized and formulated in a mathematically much more satisfactory but physically equivalent way [46,47] by resorting to stochastic dynamical equations of the Ito or Stratonowich type, ii) The model is manifestly phenomenological but it gives some clear indications concerning the fact that the macro-objectification or measurement problem admits a consistent solution, iii). The model, even though it almost completely agrees with quantum mechanics at the microscopic level qualifies itself as a rival theory of quantum mechanics, one which can be tested against it. Accordingly, it suggests where to look for putting into evidence possible violations of linearity. In recent years a lot of work in this direction is going on.

Obviously, also Collapse theories exhibit nonlocal features. However for them there is no theorem forbidding to get a generalization which does not resort to a preferred reference frame. Lot of work has been done along these lines; I will limit myself to mention some relevant steps. Before doing this, I consider it interesting to stress that the problem of having a theory inducing instantaneous reduction at-a-distance is a quite old one which has seen a lively debate and important contributions by Landau and Peierls [48], Bohr and Rosenfeld [49], Hellwig and Kraus [50] and Aharonov and Albert [51,52].

Soon after the GRW theory has been formulated, P. Pearle [53] has presented a field theoretic relativistic generalization of it which has subsequently [54] been shown to be fully Lorentz invariant. Unfortunately, the model had some limitations arising from the occurrence of divergences which were not easily amendable. In 2000 the author of the present chapter has presented [55] a genuinely relativistic toy model of a theory inducing reductions. The model satisfies all the strict conditions identified in refs.[51,52]. F. Dowker and collaborators [56] have presented, in 2004, a relativistic collapse model on a discrete space that does not require a preferred slicing of space-time.

The really important steps, however, occurred starting from 2007. R. Tumulka has presented [57] a fully satisfactory and genuinely relativistic invariant dynamical reduction model for a system of noninteracting fermions. Another important contribution [58] came from D. Bedingham. Finally, few month ago, a convincing proof of the viability of the Collapse theories in the relativistic domain has been presented [59]. The nice fact is that the conceptual attitude which underlies this attempt is that what the theory assumes to be true of the world around us is the mass density of the whole universe. In this way one recovers a unified, general picture of a quantum universe both at the micro and macro levels.

I believe that the best way to conclude this Chapter, which has dealt in detail with the compatibility of quantum effects and relativistic requirements, is to quote a clarifying sentence from R. Tumulka [57], who has studied in great detail both the Bohmian as well as the Collapse approaches to this fundamental problem:

A somewhat surprising feature of the present situation is that we seem to arrive at the following alternative: Bohmian mechanics shows that one can explain quantum mechanics, exactly and completely, if one is willing to pay with using a preferred reference slicing of space-time; our model suggests that one should be able to avoid a preferred slicing if one is willing to pay with a certain deviation from quantum mechanics.

Acknowledgements

We thank Dr. R. Romano for an accurate critical reading of the manuscript.

Author details

GianCarlo Ghirardi

Professor Emeritus, University of Triest, Italy

8. References

[1] Schrödinger E. Discussion of Probability Relations between Separated Systems. Proceedings of the Cambridge Philosophical Society 1935; 31: 555-563.

[2] Bell J.S. On the Einstein-Podolski-Rosen Paradox. Physics 1964; 1: 195-200 .

[3] von Neumann J. Matematische Grundlagen der Quantenmechanik. Berlin: Springer Verlag; 1932. (Engl. trans., Princeton University Press, 1955).

[4] Herbert N. FLASH– A Superluminal Communicator Based Upon a New Kind of Quantum Measurement. Foundations of Physics 1982; 12: 1171-79.

[5] Ghirardi GC., Rimini A. and Weber T. Unified Dynamics for Microscopic and Macroscopic Systems. Physical Review D 1986; 34: 470-491.

[6] Bell J.S. Are there Quantum Jumps? in: Schrödinger: Centenary Celebration of a Polymath. Cambridge: Cambridge University Press; 1987.

[7] Bell J.S. Against Measurement in: Sixty-two Years of Uncertainty. New York: Plenum; 1990.

[8] Bell JS. First and Second Class Difficulties in Quantum Mechanics. Journal of Physics A 2007; 40: 2921-2933.

[9] Bassi A. and Ghirardi GC. A General Argument against the Universal Validity of the Superposition Principle. Physics Letters A 2000; 275: 373-381.

[10] Bassi A. and Ghirardi GC. Dynamical Reduction Models. Physics Reports 2003; 379: 257-426.

[11] Wigner EP. Die Messung Quantenmechanischer Operatoren. Zeitschrift für Physik 1952; 131: 101-108.

[12] Araki H. and Yanase MM. Measurement of Quantum Mechanical Operators. Physical Review 1961; 120: 622-626.

[13] Yanase MM. Optimal Measuring Apparatus. Physical Review 1961; 123: 666-668.

[14] Ghirardi GC., Miglietta F., Rimini A. and Weber T. Limitations on quantum measurements. I. Determination of the Minimal Amount of Nonideality and

Identification of the Optimal Measuring Apparatuses. Physical Review D 1981; 24: 347-352.

[15] Ghirardi GC., Rimini A. and Weber T. Quantum Evolution in the Presence of Additive Conservation Laws and the Quantum Theory of Measurement. Journal of Mathematical Physics 1982; 23: 1792-1796.

[16] Kraus K. States, Effects and Operations, Berlin: Springer; 1983.

[17] Kaiser D. How the Hippies Saved Physics, New York: W.W. Norton & Co.; 2011.

[18] Cufaro Petroni N., Garuccio A., Selleri F. and Vigier JP. On a Contradiction between the Classical (idealized) Quantum Theory of Measurement and the Conservation of the Square of the Total Angular Momentum. C.R. Acad. Sci. Ser. B (Sciences Physiques)1980; 290: 111-114.

[19] Selleri F. in: International Seminar on Mathematical Theory of Dynamical Systems and Microphysics. Udine; 1979.

[20] Herbert N. QUICK-a New Superluminal Transmission Concept. Boulder Creek, Cal.: C-Life Institute; 1979.

[21] Beth R. Mechanical Detection and Measurement of the Angular Momentum of Light. Physical Review 1936; 50: 115-25.

[22] Ghirardi GC. and Weber T. On Some Recent Suggestions of Superluminal Communication through the Collapse of the Wave Function. Lettere al Nuovo Cimento 1979; 26: 599-603.

[23] Popper K. Quantum Theory and the Schism in Physics, London: Hutchinson; 1982.

[24] Ghirardi GC., Marinatto L. and de Stefano F. Critical Analysis of Popper's Experiment. Physical Review A 2007; 75: 042107-1-5.

[25] Ghirardi GC. Sneaking a Look at God's Cards, Princeton: Princeton University Press; 2005.

[26] Ghirardi GC., Rimini A. and Weber T. A General Argument against Superluminal Transmission through the Quantum Mechanical Measurement Process. Lett. Nuovo Cimento 1980; 27: 293-298.

[27] Ghirardi GC., Grassi R., Rimini A. and Weber T. Experiments of the EPR Type Involving CP-Violation do not Allow Faster-than-light Communication between Distant Observers. Europhysics Letters 1988; 6: 95-100.

[28] Eberhard PH. Bell's Theorem and the Different Concepts of Locality. Nuovo Cimento B 1978; 46: 392-419.

[29] Shimony A. Controllable and Uncontrollable Nonlocality. in: Foundations of Quantum Mechanics in the Light of New Technology. Tokyo: The physical Society of Japan; 1984.

[30] Peres A. How the No-Cloning Theorem got its Name. Fortschritte der Physik 2003; 51: 458-61.

[31] Ghirardi GC. and Weber T. Quantum Mechanics and Faster-than-Light Communication: Methodological Considerations. Il Nuovo Cimento B 1983; 78: 9-20.

[32] Wotters WK. and Zurek WH. A Single Quantum Cannot be Cloned. Nature 1982; 299: 802-803.

[33] Dieks D. Communication by EPR Devices. Physics Letters A 1982; 92: 271-272.

[34] Wigner EP. The Probability of the Existence of a Self-Reproducing Unit, in: The Logic of Personal Knowledge. London: Routledge & Kegan Paul, 1961.

[35] Alicki R. Physical Limits on Self-Replication Processes. Open Systems and Information Dynamics 2006; 13: 113-117.

[36] Eigen M. Selforganization of Matter and Evolution of Biological Macromolecules. Naturwissenschaften 1971; 58: 465-523.

[37] Greenberger DM. If one Could Build a Macroscopic Schrödinger Cat State one could Communicate Superluminally, in: Modern Studies of Basic Quantum Concepts and Phenomena. Singapore: World Scientific Publishing Co. 1998.

[38] Kalamidas DA. A Proposal for a feasible quantum-optical experiment to test the validity of the no-signaling theorem, ArXiv:1110, 4269.

[39] Ghirardi GC. and Romano R. On a Proposal of Superluminal Communication. Journal of Physics A 2012; 45: 232001.

[40] Shiekh AY. The Role of Quantum Interference in Quantum Computing. International Journal of Theoretical Physics 2006; 45: 1653-1655.

[41] Bassi A. and Ghirardi. GC. On a Recent Proposal of Faster-than-Light Quantum Communication. International Journal of Theoretical Physics 2008 47: 2500-2506.

[42] Bohm D. A Suggested Interpretation of the Quantum Theory in Terms of Hidden Variables. Physical Review 1952; 85: 166-193.

[43] Ghirardi GC. and Grassi R. Bohm's Theory versus Dynamical Reduction, in: Bohmian Mechanics and Quantum Theory, an Appraisal. The Netherlands: Kluwer Academic Publishers. 1966.

[44] Dürr, D., Goldstein, S., Münch-Berndl, K., Zanghi, N. Hypersurface Bohm-Dirac Models. Physical Review A 1999; 60: 2729-2736.

[45] Bohm, D., Hiley, B. J. The Undivided Universe. London: Routledge (1993).

[46] Pearle P. Combining Stochastic Dynamical State-Vector Reduction with Spontaneous Localization. Physical Review A 1999; 39: 2277-2289

[47] Ghirardi GC., Pearle P. and Rimini A. Markov Processes in Hilbert Space and Continuous Spontaneous Localization of Systems of Identical Particles, Physical Review A 1990; 42: 78-89.

[48] Landau and Peierls R. Erweiterung des Unbestimmtheitsprinzips für die relativistische Quantentheorie. Zeischrift für Physik, 1931; 69: 56-69.

[49] Bohr N. and Rosenfeld L. Zur Frage der Messbarkeit der Electromagnetischen Feldgrössen, Kopenaghen. 1933.

[50] Hellwig KE. and Kraus K. Formal description of measurements in local quantum field theory. Physical Review D 1970; 1: 566-571.

[51] Aharonov Y. and Albert DZ. States and observables in relativistic quantum field theories. Physical Review D 1980; 21: 3316-3324 ,

[52] Aharonov Y. and Albert DZ. Can we Make Sense out of the Measurement Process in Relativistic Quantum Mechanics? Physical Review D 1981; 24:359-370

[53] Pearle P. Toward a Relativistic Theory of Statevector Reduction, in: Sixty-Two Years of Uncertainty. Plenum Press, New York; 1990.

[54] Ghirardi GC., Grassi R. and Pearle P. Relativistic Dynamical Reduction Models: General Framework and Examples. Foundations of Physics 1990; 20: 1271-1316.

[55] Ghirardi GC. Local Measurement of Nonlocal Observables and the Relativistic Reduction Process. Foundations of Physics 2000; 38: 1337-1385.

[56] Dowker, F., Henson, J. Spontaneous Collapse Models on a Lattice. Journal of Statistical Physics 2004; 115: 1327-1339.

[57] Tumulka, R. A Relativistic Version of the Ghirardi-Rimini- Weber Model. Journal of Statistical Physics 2006; 125: 821-840.

[58] Bedingham, DJ. Relativistic state reduction dynamics. ArXiv 1003-2774v2, 2010.

[59] Bedingham DJ., Dürr D., Ghirardi GC., Goldstein S., Tumulka R. and Zanghi. N. Matter Density and Relativistic Models of Wave Function Collapse. ArXiv 1111-1425, 2012.

Quantum Information and Related Topics

Chapter 5

The Quantum Mechanics Aspect of Structural Transformations in Nanosystems

M. D. Bal'makov

Additional information is available at the end of the chapter

1. Introduction

Theoretical and experimental investigations of size effects have made a substantial contribution to the development of nanophysics and nanochemistry. However, a great deal needs to be done in this field. Experimental results are not necessarily consistent with the traditional concepts. In particular, the melting temperature of small nanoparticles unexpectedly turned out to be higher than the melting temperature of a macroscopic sample of the same chemical composition [1].

It is this chemical composition of a macroscopic system that determines its melting temperature T_m, the specific heat of melting Q_m, and the entropy of melting S_m. These quantities do not depend on the number M of atoms (in the limit, $M \to \infty$). This statement ceases to be valid for relatively small systems. In the given case, it is necessary to take into account the dependences of the quantities T_m, Q_m, and S_m on the number M of atoms. Moreover, as the size of the system decreases, the interpretation of the physical quantities T_m, Q_m, and S_m should be refined. Indeed, one molecule, for example, the hydrogen molecule, cannot melt, because its dissociation occurs with an increase in the temperature. In this respect, it is advisable to analyze the structural transformations in nanosystems within a unified approach of the first principles of quantum mechanics and statistical physics.

2. Quasiclosed ensembles

In the framework of classical physics each structural modification is set by the vector

$$\mathbf{R} = \left(\mathbf{r}_1, \mathbf{r}_2, \ldots \mathbf{r}_i, \ldots \mathbf{r}_M \right), \tag{1}$$

where r_i are the radius vectors of all atomic nuclei of the polyatomic system. But the atomic nucleus is not a mathematical point whose position is unambiguously determined by the vector r_i. The motion of microparticles is not characterized by the trajectory $r_i(t)$. One can speak solely about the sites of their localization. In the case of condensed systems the size of the sites of atomic nucleus localization is much smaller than the interatomic distances and is a tenth-hundredth of an angstrom. Therefore, one of the ways to make a brief quantum-mechanical description of the structure R consists in setting the coordinates r_i (1) of the centers of these sites.

As a rule, numerous quantum states forming a quasiclosed ensemble correspond to each memorized macrostate (to each structural modification of R_k). This raises the question about the number $G^{(0)}$ of different quasiclosed ensembles.

The magnitude of $G^{(0)}$ cannot be evaluated without application of quantum-mechanical methods. The point is that the components r_i of the vector R (1) can vary continuously, i.e. there exists a continuum of various structures (different vectors R_k), which cannot even be numbered with the help of the index k, if it has solely integer values. This hampers the determination of the number $G^{(0)}$ of different structural modifications. In a quantum-mechanical description of a structure the superfluous detailing is useless altogether since according to quantum mechanics, a system is usually localized not at one point R_k but in a certain volume (cell Ω_k). The set of all cells Ω_k is countable. It is this circumstance that allows one to speak about the number $G^{(0)}$ of different structural modifications of the condensed system with a fixed chemical composition.

In order to find the numerical value of $G^{(0)}$ it is necessary to consider primarily the problem of distribution of quantum states over different quasiclosed ensembles. Some of these are formed by the microstates corresponding to one of the free energy minima. The latter holds only for stable and metastable systems. In the overwhelming majority of cases we are dealing however the with nonequilibrium systems, the thermodynamic potentials of which are far from being extreme.

Thus structural modifications of the vitreous state are not characterized by the Gibbs energy minimum. Each of them is described by its intrinsic quasiclosed ensemble. Their macroscopic properties are invariable because a quite definite structural modification corresponds to each ensemble. It is for this reason that glasses are kinetically frozen nonequilibrium systems, the properties of which virtually do not change over the long time interval t_{max}. The same may also be said about the overwhelming majority of noncrystalline substances, many of which are already widely used for recording information.

The class of various quasiclosed ensembles (different macrostates memorized by the system with a fixed chemical composition) is extraordinarily broad. Their number is substantially larger than the number of Gibbs energy minima.

All atomic configurations R_k (1) of an ideal monatomic gas are equiprobable. Condensed systems are characterized by the totally opposite situation. Therefore, it is not surprising that some of their structural modifications may be frozen (memorized) for a long time interval t_{max}. Let us illustrate what has been said above in the framework of the adiabatic approximation.

3. Adiabatic approximation

The adiabatic approximation [2] is based on the considerable differences in the masses of electrons and nuclei, which makes it possible to describe their motions separately well. Being light particles, the electrons `succeed' in adapting themselves to the instantaneous configuration $R$ (1) of the atomic nuclei, the latter in turn `notice' only the averaged disposition of electrons.

In the zero approximation the atomic nuclei are regarded to be at rest [R = const (1)]. In this case, the wave function $\Phi_j(R, X)$ of the j-th stationary quantum state of the electron subsystem satisfies the equation [3]

$$\widehat{H}\Phi_j(R,X) = U_M^{(j)}(R)\Phi_j(R,X), \tag{2}$$

where $\widehat{H}$ is the Hamiltonian of electrons at fixed nuclei, which represents the sum of the total Coulomb energy of the interaction of atomic nuclei and electrons, the operator of the spin-orbital interaction of electrons and the operator of the kinetic energy of electrons; X is the sum of spatial and spin variables of all electrons of the system under consideration; $U_M^{(j)}(R)$ is the adiabatic electron term (Fig. 1), which in the case of a polyatomic system ($M>3$) usually has a great number of different physically non-equivalent minima R_k [4]. The Hamiltonian H does not contain any operator of the kinetic energy of atomic nuclei and, consequently, is the operator of the energy of the system under consideration for the fixed atomic configuration R.

When the motion of atomic nuclei does not induce any transitions between different electronic states, the function $U_M^{(j)}(R)$ (2) may be interpreted as the potential energy of the nuclei corresponding to the j-th electronic state. In this case their motion takes place in the potential field of $U_M^{(j)}(R)$.

Therefore, the nuclear wave function $\chi_j(R, E)$ satisfies the Schrödinger equation [3]

$$(\widehat{T} + U_M^{(j)}(R))\chi_j(R, E) = E\chi_j(R, E), \tag{3}$$

in which in contrast to (2), there is no variable X corresponding to the electron subsystem. Here, T is the operator of the kinetic energy of atomic nuclei; E is the energy of the stationary quantum state. The chemical composition n determines unequivocally the explicit form of equations (2) and (3). The components of the vector n are the relative concentrations of atoms of each species which form the system under consideration.

Their different solutions describe various modifications of a substance with a fixed composition. This can serve as the basis for classification of these solutions. Thus in the case of selenium some solutions may be attributed to the fluid state, others - to definite crystalline modifications, to amorphous modifications, to the vitreous state, to films, etc. However, it is most advisable to base the discussed classification of solutions of equations (2) and (3) on the structure R (1)

because the information about individual peculiarities of a polyatomic system is eventually stored in the mutual disposition of its atomic nuclei. Any structural modification (e.g., the k-th modification) which is preserved at least over the time interval tmax is described by the wave functions $\Phi_j(\mathbf{R}, \mathbf{X})$ (2) and $\chi_j(\mathbf{R}, E)$ (3) localized near the point $\mathbf{R}_k$. (Fig. 1). The diversity of the latter actually determines all the states belonging to the k-th quasiclosed ensemble.

Usually one or a series of potential $\mathbf{R}_k$ minima correspond to the points $U_M^{(j)}(\mathbf{R})$, near which the motion of one or other structural modification takes place. In order to estimate the number of such points (the number $G^{(0)}$ of different quasiclosed ensembles), it is, as a rule, sufficient to consider only the minima of the adiabatic electron term $U_M^{(0)}(\mathbf{R})$ corresponding to the ground ($j = 0$) state of the electron subsystem[1].

The point is that the lifetimes τ_e of most excited states of the electron subsystem are relatively short ($\tau_e \ll t_{max}$). Therefore, these states alone cannot form a quasiclosed ensemble, in the framework of which the k-th structural modification can be described over a long time interval t_{max}. Its preservation is favored by the potential barriers surrounding the minimum $\mathbf{R}_k$ of the adiabatic electron term $U_M^{(0)}(\mathbf{R})$ (Fig. 1). If they are sufficiently high, then even the low-energy quasi-steady [5] states localized in the potential well $\mathbf{R}_k$ under consideration have larger [compared to t_{max}] lifetimes τ_l which satisfy the inequalities

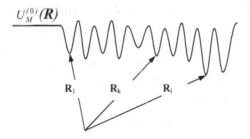

Figure 1. Adiabatic electron term $U_M^{(0)}(\mathbf{R})$. This figure is rather conditional because for polyatomic systems ($M > 3$) the function $U_M^{(0)}(\mathbf{R})$ is set, in conformity with (1), in the multidimensional space.

$$\tau_r \ll t_{max} \ll t_l \qquad (4)$$

where τ_r is the relaxation time of the phonon subsystem, which is usually appreciably shorter than the time t_{max} required for the preservation of structural modifications.

---

1 Each minimum of the function $U_M^{(0)}(\mathbf{R})$ sets one of the equilibrium configurations $\mathbf{R}_k$. Crystalline nanoparticles correspond to the deepest minima (potential wells). Most minima correspond to different noncrystalline structures. Transition from one potential well to another ($\mathbf{R}_i \to \mathbf{R}_k$) means in the general case the rearrangement of all of the M atomic nuclei of the system. The adiabatic electron term $U_M^{(0)}(\mathbf{R})$ does not depend either on temperature or on the thermal prehistory, etc. According to (2), it is unequivocally determined only by the chemical composition. Various scenarios of the system behavior consist in the sequence of passage over potential wells [the minima of the function $U_M^{(0)}(\mathbf{R})$].

Consequently, quasiclosed ensembles may be formed by the stationary and quasi-steady [5] states with large lifetimes τ_1 (4). Usually these are low-energy states, which describe vibrational motion of atomic nuclei near one of the minima of the adiabatic electron term $U_M^{(0)}$ (R).

Transitions between these states are not accompanied by any substantial changes in the $\mathbf{R}_k$ structure [6]. The k-th structural modification is preserved when only such transitions take place.

Thus, in order to preserve a polyatomic system, an exact copy and also the recorded information, it is sufficient that all changes occurring in the system do not extend outside the limits of one and the same quasiclosed ensemble. It is this ensemble that characterizes the properties of the system displayed during informational interaction [7].

The magnitude of $G^{(0)}$. can be estimated proceeding from the number $J(\mathbf{n}, M)$ of different minima of the potential $U_M^{(0)}$ (R). This approach allows a relatively simple derivation of numerical estimates as the function $J(\mathbf{n}, M)$ depends on only two arguments and, in addition, its determination is actually based on equation (2) when j= 0. This unambiguous mathematical definition is useful not only for the problem of information copying and recording but also for considering a wide range of other issues [4].

4. Estimation of the number $G^{(0)}$ of different quasiclosed ensembles

For the number $J(\mathbf{n}, M)$ of different physically non-equivalent local minima of the adiabatic electron term $U_M^{(0)}$ (R), which corresponds to the ground electronic state of the electroneutral system consisting of M atoms, the following asymptotic formula [4] is valid as $M \to \infty$

$$\frac{1}{M} lnJ(\mathbf{n}, M) \sim \alpha_\mathbf{n} ,$$ (5)

where $\alpha_\mathbf{n}$ is the positive parameter dependent solely on the chemical composition **n.** The components n_i of the vector **n** are the relative concentrations of atoms of each type.

It follows from (5) that

$$J(\mathbf{n}, M) = exp(\alpha_\mathbf{n} M + o(M)),$$ (6)

the function $o(M)$ satisfying the condition $\lim_{M \to \infty} o(M)/M = 0$. In other words, the number $J(\mathbf{n}, M)$ of different physically nonequivalent minima of the $U_M^{(0)}$ (R) potential exhibits a rapid exponential growth with the increasing number M of atoms forming the system with a fixed (n = const) chemical composition. The numerical values of $\alpha_\mathbf{n}$ in relationship (6) usually differ from $\ln 2 \approx 0.69$ by smaller than one order of magnitude.

This fact is not surprising because the magnitude $J(\mathbf{n}, M)$ (6) takes into account all potentially possible structural modifications $\mathbf{R}_k$ (1) of a polyatomic system. These are structures of nanoparticles, liquid, glass, perfect crystal, crystals with different concentrations of particular defects, polycrystals, amorphous substances, amorphous and vitreous films, glass-ceramics and many others, including the structures of microheterogeneous materials storing the recorded information. These structures differ from each other not only in the location of particular defects, holes, etc. There exist other differences. For example, there are six equilibrium positions for each oxygen atom in the structure of β cristobalite (Fig. 2, positions 1–6) [6].

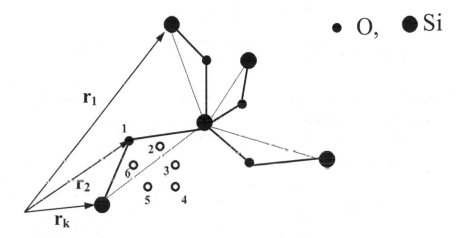

Figure 2. Structural fragment of β cristobalite (r_i are the radius vectors of equilibrium atomic positions).

The transitions between them are not accompanied by the formation (disappearance) of defects, holes, dangling chemical bonds, etc. Each structure thus formed is not an exact copy of other structures [7]. These configurations are also taken into account by relationship (6).

The diversity of elementary configurational excitations particularly involves small structural transformations. As a result of these transformations, the transition from one minimum of the adiabatic electronic term $U_M^{(0)}$ (**R**) to another minimum occurs through a correlated rearrangement of many atoms involved in a particular nanofragment. In this case, the distances between any pair of neighboring atoms change insignificantly as compared to the interatomic distances (Fig. 3) and, as a consequence, the short-range order is retained.

Specifically, small structural transformations occur in the glass transition range [4] (upon softening of a glass and melting of a crystal) when the coordination numbers remain virtually unchanged. Uncorrelated small structural transformations that proceed in different nanofragments of the melt upon its rapid cooling lead to generation of internal stresses in the resulting glass.

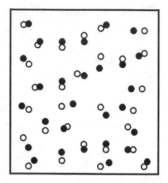

Figure 3. Schematic diagram illustrating a small structural transformation: ○initial and ●final positions of sites of atomic nuclei in a nanofragment.

The diversity of minima of the function $U_M^{(0)}(\mathbf{R})$ makes it possible to explain the possibility to vary properties of a material of the same chemical composition through preparation of its various modifications described by different quasiclosed ensembles.

Number $G^{(0)}$ of different quasiclosed ensembles satisfies the following relations :

$$ ln\,G^{(0)}(t_{max},\, n,\, M) \le ln J\,(n,\, M) = (\alpha_n M + o(M)) \le (B \times M + o(M)), \tag{7} $$

where constant B is determined by the identity

$$ B \equiv \sup_n \alpha_n \tag{8} $$

It would not be particularly difficult to find the exact value of constant B (8) if the solutions of equation (2) at j=0 were known for the systems of various chemical compositions. Since this is not the case, one has to use model approaches. In their frameworks it is possible to calculate numerical values of the parameter α_n (5), (8) for specific systems. The results of such computations [4] support the following estimate:

$$ B \approx 3 \tag{9} $$

It is difficult to investigate thoroughly all configurations of the polyatomic system, because their number is exponentially large [see relationship (6)]. In this respect, it is necessary to use model approaches. A model based on the Gaussian distribution is convenient for constructing the statistical thermodynamics of melting and softening of nanoparticles.

5. Model spectrum for the description of configurational excitations

In order to describe any equilibrium process, including the melting, in the framework of statistical thermodynamics [8], it is sufficient to know the time dependence of the statistical sum Z, which is uniquely determined by the temperature T, the energies E_j of stationary quantum states, and the multiplicity g_j of their degeneracy; that is

$$Z = \sum_j g_j \, exp(-E_j \,/\, kT). \qquad (10)$$

Here, k is the Boltzmann constant and $E_j < E_{j+1}$. This holds true for any system with a fixed number of particles from one elementary particle to inhomogeneous (specifically, multiphase) media.

The energy spectrum $\{E_j, g_j\}$ depends on the number M of atoms. Since the energy spectrum of a macroscopic system ($M \geq 10^{20}$) differs from that of a nanoparticle ($10^8 \geq M \geq 10$), the processes accompanying the melting of a macroscopic sample and a nanoparticle of the same chemical composition cannot not be completely identical. However, the specific features of these processes have much in common: in both cases, upon melting, the structure undergoes transformations, the system becomes microscopically labile, and the entropy and the internal energy increase abruptly.

In expression (10), the summation is performed over all possible configurations corresponding to relationship (6) and different vibrational states. By using the known analytical expression for the statistical sum of an oscillator [8], the sum of all terms associated with the i-th configuration $\mathbf{R}_i$ can be represented in the form $\tilde{g}_i exp(E/kT)$.

This enables us to change over to the model partition function with due regard only for the configurations in which each configuration is included $\tilde{g}_i/\tilde{g}_1$ times, where $\tilde{g}_1$ corresponds to the configuration with a minimum energy[2]. The ratios $\tilde{g}_i/\tilde{g}_1$ actually take into account the role of the phonon subsystem in the melting.

The energies $E_i = U_M^{(i)}(\mathbf{R}_i)$ of the equilibrium configurations (2) of the polyatomic system are conveniently calculated per atom; that is,

$$\varepsilon_i = E_i \,/\, M \qquad (11)$$

The spectrum of numerical values of the energies ε_i (fig.4) depends on the number of atoms M. The level with the minimum energy $E_1 = 0$ is assumed to be nondegenerate: $g_1 = 1$. The distribution of levels located in the energy range between $M\varepsilon_g$ and $M\varepsilon_c$ is approximated by the Gaussian distribution

---

2 In this case, the number of configurations can be determined from formula (6), in which the numerical value of the parameter α_n changes insignificantly.

$$N(E) \sim \frac{\sigma(E-E_0)}{\gamma\sqrt{2\pi M}} exp\left\{M\alpha_n - 0.5(\frac{E-M\ h}{\gamma\sqrt{M}})^2\right\} \tag{12}$$

Here, E_0 is the minimum energy necessary for transforming the system from the ground state into the excited state, $\sigma(E - E_0)$ is the step function,

$$\gamma = 0.5(\varepsilon_c - \varepsilon_g)/(2\alpha_n)^{1/2} \tag{13}$$

$$h = 0.5\,(\varepsilon_g + \varepsilon_c). \tag{14}$$

From the distribution density $N(E)$ (12), we derive the following analytical expression for the statistical sum (10) (statistical integral)

$$Z = 1 + \left\{expM\,(\alpha_n - \frac{h}{kT} + \frac{0.5\gamma^2}{(kT)^2})\right\}\Phi(\frac{M\ h-E_0}{\gamma\sqrt{M}} - \frac{\gamma\sqrt{M}}{kT}) \Big/ \Phi(\frac{M\ h-E_0}{\gamma\sqrt{M}}), \tag{15}$$

where $\Phi(x) = \frac{1}{\sqrt{2\pi}}\int_{-\infty}^{x} exp(-0.5x^2)dx$ is the normal distribution function. The analytical expression (15) permits us to analyze not only the dependence of the statistical sum on the number M of atoms. The corresponding dependences can be obtained for all quantities that are uniquely defined by the statistical sum Z. In particular, these are the melting temperature, the heat of melting, and the entropy of melting.

In the limit $M \rightarrow \infty$, the melting temperature T_m is given by the expression [4]

$$T_m = (\sqrt{\varepsilon_g} + \sqrt{\varepsilon_c})^2\,/\,4k\alpha_n \tag{16}$$

In the same limit, the stepwise increments of the energy $\Delta\varepsilon$ (fig.4) and the entropy Δs upon melting per atom are represented by the formulas

$$\Delta\varepsilon = (\varepsilon_g\varepsilon_c)^{1/2}, \tag{17}$$

$$\Delta s = 4k\alpha_n(e_g e_c)^{1/2}(\sqrt{\varepsilon_g} + \sqrt{\varepsilon_c})^{-2} \leq k\alpha_n. \tag{18}$$

It should be noted that, at the melting temperature T_m, the following inequalities are satisfied:

$$\varepsilon_c \ / \ k\alpha_n \geq T_m \geq max\{ \ (\varepsilon_g \varepsilon_c)^{1/2} \ / \ k\alpha_n, \ \ \varepsilon_c \ / \ 4k\alpha_n \ \}. \tag{19}$$

Before melting, the energy is minimum. Without loss of generality, this energy can be taken equal to zero. In the course of melting, there occurs a stepwise transition within the energy band $[\varepsilon_g, \varepsilon_c]$, which involves energies of the majority of the equilibrium configurations (fig.4). The width $(\varepsilon_c - \varepsilon_g)$ of this band is proportional to the root-mean-square deviation of the numerical values of the energies ε of different configurations.

Upon melting, the structure undergoes transformations. Furthermore, the nanoparticle becomes labile. In particular, the nanoparticle changes in shape, because, after melting, there occur spontaneous transitions between the structural modifications with close energies in the energy band $[\varepsilon_g, \varepsilon_c]$ (fig.4).

The notion of the "melting of nanoclusters" has already been used [1]. It is obvious that the processes accompanying the melting of a macroscopic sample and a nanoparticle cannot not be completely identical. However, the specific features of these processes have much in common. In both cases, upon melting, the structure undergoes transformations, the system becomes labile, and the entropy and the internal energy increase abruptly.

According to relationships (16-18), the melting temperature T_m can be described by one more expression

$$T_m = \Delta\varepsilon \ / \ \Delta s, \tag{20}$$

which coincides with the known expression that relates the heat $\Delta\varepsilon$, the entropy Δs, and the temperature T_m of the transition [9]. Note that, in the case of the macroscopic system, the heat of melting $\Delta\varepsilon$ and the entropy of melting Δs per atom in relationship (20) are independent of the number M of atoms forming the macroscopic system, whereas the opposite situation is observed for the nanoparticle. The spectrum of energies ε of equilibrium configurations can even radically change (fig.4).

For example, the two-atom system ($M = 2$) has only one equilibrium configuration $J(\mathbf{n}, 2) = 1$ and the energy band $[\varepsilon_g, \varepsilon_c]$ is absent. Since the structure should change upon melting (there should occur a transition from one equilibrium configuration to another equilibrium configuration), the melting of two-atom systems, in principle, is impossible.

However, the above concept is inapplicable to relatively small nanoclusters consisting of 13 atoms with $J(\mathbf{n}, 13) = 1478$ different configurations (different energy levels) [10]. In this case, the energy spectrum can be described by the Gaussian distribution but with parameters different from those used for the macroscopic system.

Relationships (16)–(20) are also valid in the mesoregion where the number of atoms M is larger than two but is not sufficient for the applicability of the asymptotic relationship (5), which allows one to estimate the number of equilibrium configurations of the macroscopic system.

For nanoclusters, relationships (16) and (18) should contain the parameter $\tilde{\alpha} = \{\ln J(n, M)\text{-}1\} / M$ instead of the parameter α_n involved in relationship (5). In the case of a two-atom system, we have the parameter $\tilde{\alpha} = \{\ln(J(n, 2) - 1)\}/2 = -\infty$. For a nanocluster consisting of 13 atoms, we should use the parameter $\tilde{\alpha} = \{\ln(J(n, 13) - 1)\}/13 \approx 0{,}56$. The change in sign of the parameter $\tilde{\alpha}$ indicates that, for a relatively small number of atoms M, which satisfies the inequality $M > 2$, the parameter $\tilde{\alpha}$ can turn out to be close to zero.

Therefore, the numerical value of the parameter in the mesoregion can appear to be considerably smaller than the parameter α_n involved in relationship (5) and used for calculating the melting temperature of the macroscopic sample according to relationship (16). This circumstance is responsible for the observed increase in the melting temperature of sufficiently small nanoparticles as compared to the macroscopic sample.

Since the parameter $\alpha_n(\tilde{\alpha})$ is equal to the natural logarithm of the number of energy levels (equilibrium configurations) in the energy band $[\varepsilon_g, \varepsilon_c]$ (fig.4), the product $k\alpha_n$ in relationship (18) gives an estimate from above for the jump Δs in the configurational entropy upon melting. In the case where $\alpha_n(\tilde{\alpha}) \to 0$, we have $\Delta s \to 0$. In other words, the decrease in the parameter $\alpha_n(\tilde{\alpha})$, according to relationship (18), leads to a decrease in the entropy of melting Δs and, consequently, to an increase in the melting temperature T_m in accordance with relationship (20).

Therefore, generally speaking, the melting temperature of macroscopic samples can be lower than the melting temperature of nanoclusters of the same chemical composition. Moreover, there are other specific features of melting of nanoparticles. Particularly, this refers to the melting temperature range ΔT_m, which, unlike the corresponding range for macroscopic systems, is not a small quantity.

6. On the temperature ranges of melting and softening

A decrease in the number of atoms M results in an increase in the temperature range ΔT_m of the phase transition [10]; that is,

$$\Delta T_m \approx T_m \left(2nln10\right) \ / \ \left(M\tilde{\alpha}\right). \tag{21}$$

Here, n is a natural number which, at the boundaries of this range, determines a low probability $p = 10^{-n}$ that the system is in the liquid state before melting and in the solid state after melting, respectively.

At $n = 3$ and $\tilde{\alpha} = \ln2$, expression (21) can be represented in the form

$$\frac{\Delta T_m}{T_m} M \approx 20. \tag{22}$$

As follows from formula (22), the quantities ΔT_m and T_m for nanoparticles containing of the order of ten atoms are comparable in magnitude. By contrast, the temperature range ΔT_m for macroscopic systems ($M \sim 10^{23}$) is nearly equal to zero.

According to relationship (22), the temperature range ΔT_m is relatively small

$$\frac{\Delta T_m}{T_m} \ll 10^{-2}. \tag{23}$$

only for systems involving a considerable number of atoms

$$M \gg 2000. \tag{24}$$

Otherwise, the quantity ΔT_m should not be ignored. Therefore, specific analogy can be drawn between the melting of nanoparticles and the softening of glasses.

Actually, the microscopic mechanism of glass softening is associated with the independent structural excitations in medium-range order nanofragments. Their initial structures, as a rule, are not exact copies of each other [7]. As a consequence, since the glass softening is a thermodynamically nonequilibrium irreversible process, it occurs in a specific temperature range ΔT_g rather than at a fixed temperature T_g.

Therefore, the glass softening and the transition of the nanoparticle to the microscopically labile state proceed in a particular temperature range rather than at a fixed temperature. Both these phenomena are responsible for the inelastic compliance of the system. This manifests itself as a viscous flow for macroscopic systems and a possibility of changing the shape due to the spontaneous transitions between different structural modifications with close energies within the band $[M\varepsilon_g, M\varepsilon_c]$ (Fig. 4) for nanoparticles.

The transition of the nanoparticle to the microscopically decrease in the temperature, the nanoparticle structure does not always revert to the initial state and, as in the case of the glass transition, one of the intermediate structures can turn out to be frozen. The question arises of whether the transition of the nanoparticle to the microscopically labile state in similar situations can be always interpreted as softening.

7. Admissible states

The freezing is a thermodynamically nonequilibrium process. The concept of "admissible states" [8] is useful when constructing the statistical thermodynamics of these processes. Not all states can occur for the observation time of a specific system. The states in which SiO_2 has a crystalline form are inadmissible at low temperatures if the object was initially in the vitreous form: this compound in experiments at low temperatures does not transform into quartz

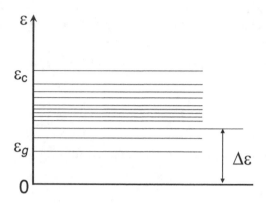

Figure 4. Spectrum of energies ε per atom for equilibrium configurations of the polyatomic system.

during our life. We assume that all quantum states are admissible if they are not excluded according to the definition of the system or the chosen time scale [8].

Generally speaking, the spectrum of admissible states changes depending on the prehistory of the formation of the polyatomic system. The same holds true for numerical values of the parameters γ, h, α_n, ε_c, and ε_g (12)–(14) used for approximating the spectrum of admissible states. Let us assume that the initial state of the nanoparticle is in the band $[M\varepsilon_g, M\varepsilon_c]$ (Fig. 4) and the state with the minimum energy $\varepsilon = 0$ is inadmissible[3]. Then, the spectrum of admissible states is approximated only by distribution (12). In this case, we have $\varepsilon_g = 0$ and the heat of melting $\Delta\varepsilon$ according to relationship (17) is equal to zero. As a consequence, we cannot speak about the melting, even though the thermodynamically nonequilibrium transition to the microscopically labile state is possible upon heating. This transition occurs in the temperature range ΔT_g at the softening temperature T_g, which can be estimated as follows [4]:

$$T_g \sim \gamma / (k(2\alpha_n)^{1/2}) \tag{25}$$

The softening temperature T_g and the boundaries of the softening range ΔT_g are kinetic parameters and depend on the prehistory of the compound. By contrast, the melting is a thermodynamically equilibrium process, which proceeds at a fixed temperature. As a result of melting, the structure of the macroscopic system is radically changed from crystalline to disordered. Upon softening, the short-range order in the atomic arrangement is retained. Consequently, the structure undergoes an insignificant transformation. Correspondingly, a change in the internal energy is also small and, therefore, the notion of "the heat of softening" does not exist.

---

3 The lower level corresponds to the crystal. This level is excluded when constructing the statistical thermodynamics of glasses and glass-forming melts. This level is not used for describing the softening and glass transition.

In the case of macroscopic systems, the above criteria allow us to distinguish rather simply the melting from softening. By contrast, not all transitions of the nanoparticle from the solid state to the microscopically labile state can be uniquely interpreted as melting or softening, because there are intermediate situations.

Specifically, these situations involve a thermodynamically equilibrium transition that results in an insignificant change in the structure (the short-range order is retained). In this case, the jump $\Delta\varepsilon$ (17) in the internal energy

$$\Delta\varepsilon << 10^{-2} eV \tag{26}$$

is small as compared to the heat of melting of the macroscopic system of the same chemical composition per atom. It is unlikely that this transition should be treated as melting. However, since the transition under consideration is thermodynamically equilibrium, it is not advisable to identify this transition with the softening.

Eventually, it is important to know the spectrum of admissible states and the parameters γ, h, α_n, ε_c, and ε_g (12)–(14). This makes it possible to reveal the energy characteristics of elementary structural excitations of the nanoparticle. In particular, the transitions between the equilibrium configurations closest in energy in the band $[M\varepsilon_g, M\varepsilon_c]$ (Fig. 1) are accompanied by the absorption (emission) of longwavelength photons [4]. Their frequencies can be estimated from the relationship

$$\nu[Hz] \sim 2 \cdot 10^{14} M\left(\varepsilon_c - \varepsilon_g\right) exp\left(-\alpha_n M\right). \tag{27}$$

According to relationship (27), at $M = 100$, $\alpha_n = 0.1$, and $(\varepsilon_c - \varepsilon_g) = 0.1$ eV, we obtain a frequency $\sim 10^{11}$ Hz, which corresponds to the microwave range of electromagnetic radiation.

It follows from relationship (27) that the elementary structural excitations of nanoparticles can be attended by emission (absorption) of photons with frequencies in the microwave, radiofrequency, and low-frequency ranges. It was experimentally demonstrated that the microwave radiation can accelerate chemical reactions by a factor of several tens and even several hundreds [11]. The microscopic mechanism of this phenomenon is not clearly understood. However, it is unquestionable that this mechanism is not reduced only to heating.

8. Conclusions

A detailed (on the microscopic level) analysis of the processes that occur upon transition of nanoparticles to the microscopically labile state stimulates consideration of a number of fundamental problems. Their solution provides a deeper insight into the specific features of the nanoworld. Indeed, the melting and softening cannot proceed in the absence of an exponentially large number of various structural modifications (6). However, up to now, most attention has been focused on relatively stable structures. The number of these structures for a nanocluster composed of 13 atoms is considerably smaller than 1478 [10].

The other structural modifications have not been adequately investigated, even though their role is important not only for the transition of the nanoparticle to the microscopically labile state. Many chemical transformations represent a sequence of transitions between unstable modifications. They should be taken into account when developing methods for synthesizing nanostructured functional materials with controlled properties.

The majority of nanoparticles of the same chemical composition exhibit similar additive properties. It is sufficient to investigate one of these nanoparticles in order to judge the properties of the other nanoparticles. However, there are "special" nanoparticles. Their properties differ noticeably from the statistical-mean properties and can be unique as compared to those of macromolecules and compact materials. Owing to this uniqueness, it is these nanoparticles that are of most interest for the nanotechnology.

Certainly, special nanoparticles are small in number. Among an exponentially large number (6) of various nanoparticles of the same chemical composition, the choice of a special nanoparticle with required properties is not a simple problem. Moreover, it is not a priori known whether there exists this nanoparticle in principle.

Furthermore, the potential possibility of occurring a large number of similar structural modifications different from the required modification complicates the reproduction of an exact copy [7] of the nanoparticle under consideration. That is why the reproducibility is one of the key problems of the nanotechnology.

In actual practice, the special nanoparticle cannot be synthesized in an accidental way. The traditional methods are not necessarily effective because the vast majority of the currently used chemical reactions belong to "disorganized" reactions in which particles (molecules, ions, atoms, radicals) react as a result of random collisions.

In order to solve many problems of nanotechnologies, it is required to control chemical processes on the microscopic level. It is necessary to design nontraditional methods based on nonequilibrium processes [12].

In particular, it seems likely that the use of electromagnetic radiation holds considerable promise. The methods of microwave chemistry have already been used to produce nanopowders [11].

The problem associated with the synthesis of special nanoparticles would be completely solved if the technique for preparing any controlled equilibrium configuration R_i of atomic nuclei (Fig. 1) would be developed. In general, for this purpose, it is necessary to know how to operate not with one atom but with many atoms according to a special program [13], which represents algorithmic information. It should be noted that the information aspect of microscopic processes [14] has come under the scrutiny of science only in recent years.

Author details

M. D. Bal'makov*

Address all correspondence to: Balmak1@yandex.ru

Faculty of Chemistry, St. Petersburg State University, Staryi Peterhof, Universitetskii Pr. 26, St. Petersburg, Russia

References

[1] Shvartsburg, A. A, & Jarrold, M. F. Solid Clusters Above the Bulk Melting Point, Phys. Rev. Lett., (2000). , 85(12), 2530-2532.

[2] Born, M, & Oppenheimer, J. R. Zur Quantentheorie der Molekeln, Ann. Phys. (Leipzig), (1927). , 84(20), 457-483.

[3] Nikitin, E. E. Teoriya elementarnykh atomno-molekulyarnykh protsessov v gazakh (The Theory of Elementary Atomic-Molecular Processes in Gases), Moscow: Khimiya, (1970). , 1-456.

[4] Bal'makov M.D., Stekloobraznoe sostoyanie veshchetsva (The Vitreous State of Matter), St. Petersburg: St. Petersburg State University, (1996). , 1-184.

[5] Landau, L. D, & Lifshitz, E. M. Kvantovaya mekhanika. Nerelyativistskaya teoriya (Quantum Mechanics: Nonrelativistic Theory), Moscow: Fizmatgiz, (1963). , 1-704.

[6] Wright, A. F, & Leadbetter, A. J. The Structure of the β-Cristobalite Phases of SiO_2 and $AlPO_4$, Philos. Mag., (1975). , 31(6), 1391-1401.

[7] Bal'makov M.D., Information Capacity of Condensed Systems, Semiconductors and semimetals. Elsevier academic press, (2004). , 79, 1-14.

[8] Kitel Ch., Elementary Statistical Physics, New York: Wiley, 1958. Translated under the title Statisticheskaya termodinamika, Moscow: Nauka, (1977). , 1-336.

[9] Landau, L. D, & Lifshitz, E. M. Statisticheskaya fizika Moscow: Nauka, (1964). Translated under the title Course of Theoretical Physics, Statistical Physics, Oxford: Butterworth-Heinemann, 1968., 5, 1-568.

[10] Bal'makov M.D., On the Melting Temperature of Nanoparticles, Fiz. Khim. Stekla, (2008). Glass Phys. Chem. (Engl. transl.), 2008, vol. 34, no. 1, pp. 110-112]., 34(1), 140-143.

[11] Berdonosov, S. S. Microwave Chemistry, Soros. Obraz. Zh., (2001). , 7(1), 32-38.

[12] Tver'yanovich A.S., Kim, D., Borisov, E.N., et al., Lazernye i mikrovolnovye metody polucheniya i modifikatsii khal'kogenidnykh poluprovodnikovykh materialov (Laser

and Microwave Methods for Preparing Chalcogenide Semiconductor Materials), St. Petersburg: BKhV, (2006). , 1-99.

[13] Bal'makov M.D., Algorithmic Approach to the Problem of Control of the Structure of the Nanostate of Matter, Vestn. St. Peterb. Univ., Ser. 4: Fiz., Khim., (2005). (2), 51-59.

[14] Kadomtsev, B. B. Dinamika i informatsiya (Dynamics and Information), Moscow : Usp. Fiz. Nauk Publishing,(1997). in Russian].

Chapter 6

Decoding the Building Blocks of Life from the Perspective of Quantum Information

Rodolfo O. Esquivel, Moyocoyani Molina-Espíritu,
Frank Salas, Catalina Soriano, Carolina Barrientos,
Jesús S. Dehesa and José A. Dobado

Additional information is available at the end of the chapter

1. Introduction

Physical theories often start out as theories which only embrace essential features of the macroscopic world, where their predictions depend on certain parameters that have to be either assumed or taken from experiments; as a result these parameters cannot be predicted by such theories. To understand why the parameters have the values they do, we have to go one level deeper—typically to smaller scales where the easiest processes to study are the ones at the lowest level. When the deeper level reduces the number of unknown parameters, we consider the theory to be complete and satisfactory. The level below conventional molecular biology is spanned by atomic and molecular structure and by quantum dynamics. However, it is also true that at the lowest level it becomes very difficult to grasp all the features of the molecular processes that occur in living systems such that the complexity of the numerous parameters that are involved make the endeavour a very intricate one. Information theory provides a powerful framework for extracting essential features of complicated processes of life, and then analyzing them in a systematic manner. In connection to the latter, quantum information biology is a new field of scientific inquiry in which information-theoretical tools and concepts are permitting to get insight into some of the most basic and yet unsolved questions of molecular biology.

Chirality is often glossed over in theoretical or experimental discussions concerning the origin of life, but the ubiquity of homochiral building blocks in known biological systems demands explanation. Information theory can provide a quantitative framework for understanding the role of chirality in biology. So far it has been thought that the genetic code is "unknowable" by considering DNA as a string of letters only (... ATTGCAAGC...) and likewise by considering proteins as strings of identifiers (... DYRFQ...), we believe that this particular conclusion might be probably wrong because it entirely fails to consider the information content of the molecular structures themselves and their conformations.

On the other hand, according to molecular biology, living systems consist of building blocks which are encoded in nucleic acids (DNA and RNA) and proteins, which possess complex patterns that control all biological functions. Despite the fact that natural processes select particular building blocks which possess chemical simplicity (for easy availability and quick synthesis) and functional ability (for implementing the desired tasks), the most intriguing question resides in the amino acid selectivity towards a specific codon/anticodon. The universal triplet genetic code has considerable and non-uniform degeneracy, with 64 codons carrying 21 signals (including Stop) as shown in Table 1. Although there is a rough rule of similar codons for similar amino acids, no clear pattern is obvious.

Information theory of quantum many-body systems is at the borderline of the development of physical sciences, in which major areas of research are interconnected, i.e., physics, mathematics, chemistry, and biology. Therefore, there is an inherent interest for applying theoretic-information ideas and methodologies to chemical, mesoscopic and biological systems along with the processes they exert. On the other hand, in recent years there has been an increasing interest in applying complexity concepts to study physical, chemical and biological phenomena. Complexity measures are understood as general indicators of pattern, structure, and correlation in systems or processes. Several alternative mathematical notions have been proposed for quantifying the concepts of complexity and information, including the Kolmogorov–Chaitin or algorithmic information theory (Kolmogorov, 1965; Chaitin, 1966), the classical information theory of Shannon and Weaver (Shannon & Weaver, 1948), Fisher information (Fisher, 1925; Frieden, 2004), and the logical (Bennet, 1988) and the thermodynamical (Lloyd & Pagels, 1988) depths, among others. Some of them share rigorous connections with others as well as with Bayes and information theory (Vitanyi & Li, 2000). The term complexity has been applied with different meanings: algorithmic, geometrical, computational, stochastic, effective, statistical, and structural among others and it has been employed in many fields: dynamical systems, disordered systems, spatial patterns, language, multielectronic systems, cellular automata, neuronal networks, self-organization, DNA analyses, social sciences, among others (Shalizi et al., 2004; Rosso et al., 2003; Chatzisavvas et al., 2005; Borgoo et al., 2007).

The definition of complexity is not unique, its quantitative characterization has been an important subject of research and it has received considerable attention (Feldman & Crutchfield, 1998; Lamberti et al., 2004). The usefulness of each definition depends on the type of system or process under study, the level of the description, and the scale of the interactions among either elementary particles, atoms, molecules, biological systems, etc.. Fundamental concepts such as uncertainty or randomness are frequently employed in the definitions of complexity, although some other concepts like clustering, order, localization or organization might be also important for characterizing the complexity of systems or processes. It is not clear how the aforementioned concepts might intervene in the definitions so as to quantitatively assess the complexity of the system. However, recent proposals have formulated this quantity as a product of two factors, taking into account order/disequilibrium and delocalization/uncertainty. This is the case of the definition of López-Mancini-Calbet (LMC) shape complexity [9-12] that, like others, satisfies the boundary conditions by reaching its minimal value in the extreme ordered and disordered limits. The LMC complexity measure has been criticized (Anteonodo & Plastino, 1996), modified (Catalán et al., 2002; Martin et al., 2003) and generalized (López-Ruiz, 2005) leading to a useful estimator which satisfies several desirable properties of invariance under scaling transfromations, translation, and replication (Yamano, 2004; Yamano, 1995). The utility of this improved complexity has been verified in many fields [8] and allows reliable detection of periodic, quasiperiodic, linear stochastic, and chaotic dynamics (Yamano, 2004; López-Ruiz et al., 1995; Yamano, 1995). The LMC measure is constructed as the product of two important information-theoretic quantities (see below): the so-called disequilibrium D (also known as self-similarity (Carbó-Dorca et al., 1980) or information energy Onicescu, 1996), which quantifies the departure of the probability density from uniformity (Catalán et al., 2002; Martinet al., 2003) (equiprobability) and the Shannon entropy S, which is a general measure of randomness/uncertainty of the probability density (Shannon & Weaver, 1948), and quantifies the departure of the probability density from localizability. Both global quantities are closely related to the measure of spread of a probability distribution.

The Fisher-Shannon product FS has been employed as a measure of atomic correlation (Romera & Dehesa, 2004) and also defined as a statistical complexity measure (Angulo et al., 2008a; Sen et al., 2007a). The product of the power entropy J -explicitly defined in terms of the Shannon entropy (see below)- and the Fisher information measure, I, combine both the global character (depending on the distribution as a whole) and the local one (in terms of the gradient of the distribution), to preserve the general complexity properties. As compared to the LMC complexity, aside of the explicit dependence on the Shannon entropy which serves to measure the uncertainty (localizability) of the distribution, the Fisher-Shannon complexity replaces the disequilibrium global factor D by the Fisher local one to quantify the departure of the probability density from disorder (Fisher, 1925; Frieden, 2004) of a given system through the gradient of the distribution.

The Fisher information I itself plays a fundamental role in different physical problems, such as the derivation of the non-relativistic quantum-mechanical equations by means of the minimum I principle (Fisher, 1925; Frieden, 2004), as well as the time-independent Kohn-Sham equations and the time-dependent Euler equation (Nagy, 2003; Nalewajski, 2003). More recently, the Fisher information has been employed also as an intrinsic accuracy measure for specific atomic models and densities (Nagy & Sen, 2006; Sen et al., 2007b)), as well as for general quantum-mechanical central potentials (Romera et al. 2006; Dehesa et al., 2007). The concept of phase-space Fisher information has been analyzed for hydrogenlike atoms and the isotropic harmonic oscillator (Hornyak & Nagy, 2007), where both position and momentum variables are included. Several applications concern atomic distributions in position and momentum spaces have been performed where the FS complexity is shown to provide relevant information on atomic shell structure and ionization processes (Angulo et al., 2008a; Sen et al., 2007a; Angulo & Antolín, 2008b; Antolín & Angulo, 2009).

In line with the aforementioned developments we have undertaken multidisciplinary research projects so as to employ IT at different levels, classical (Shannon, Fisher, complexity, etc) and quantum (von Neumann and other entanglement measures) on a variety of chemical processes, organic and nanostructured molecules. Recently, significant advances in chemistry have been achieved by use of Shannon entropies through the localized/delocalized features of the electron distributions allowing a phenomenological description of the course of elementary chemical reactions by revealing important chemical regions that are not present in the energy profile such as the ones in which bond forming and bond breaking occur (Esquivel et al., 2009). Further, the synchronous reaction mechanism of a S_N2 type chemical reaction and the non-synchronous mechanistic behavior of the simplest hydrogenic abstraction reaction were predicted by use of Shannon entropies analysis (Esquivel et al., 2010a). In addition, a recent study on the three-center insertion reaction of silylene has shown that the information-theoretical measures provide evidence to support the concept of a continuum of transient of Zewail and Polanyi for the transition state rather than a single state, which is also in agreement with other analyses (Esquivel et al., 2010b). While the Shannon entropy has remained the major tool in IT, there have been numerous applications of Fisher information through the "narrowness/disorder" features of electron densities in conjugated spaces. Thus, in chemical reactions the Fisher measure has been employed to analyze its local features (Esquivel et al., 2010c) and also to study the steric effect of the conformational barrier of ethane (Esquivel et al., 2011a). Complexity of the physical, chemical and biological systems is a topic of great contemporary interest. The quantification of complexity of real systems is a formidable task, although various single and composite information-theoretic measures have been proposed. For instance, Shannon entropy (S) and the Fisher information measure (I) of the probability distributions are becoming increasingly important tools of scientific analysis in a variety of disciplines. Overall, these studies suggest that both S and I can be used as complementary tools to describe the information behavior, pattern, or complexity of physical and chemical systems and the electronic processes involving them. Besides, the disequilibrium (D), defined as the expectation value of the probability density is yet another complementary tool to study complexity since it measures its departure from equiprobability. Thus, measuring the complexity of atoms and molecules represents an interesting area of contemporary research which has roots in information theory (Angulo et al., 2010d). In particular, complexity measures defined as products of S and D or S and I have proven useful to analyze complexity features such as order, uncertainty and pattern of molecular systems (Esquivel et al., 2010f) and chemical processes (Esquivel et al., 2011b). On the other hand, the most interesting technological implications of quantum mechanics are based on the notion of entanglement, which is the essential ingredient for the technological implementations that are foreseen in the XXI century. Up to now it remains an open question whether entanglement can be realized with molecules or not and hence it is evident that the new quantum techniques enter the sphere of chemical interest. Generally speaking, entanglement shows up in cases where a former unit dissociates into simpler sub-systems, the corresponding processes are known quite well in chemistry. Although information entropies have been employed in quantum chemistry, applications of entanglement measures in chemical systems are very scarce. Recently, von Neumann measures in Hilbert space have been proposed and applied to small chemical systems (Carrera et al

2010, Flores-Gallegos and Esquivel, 2008), showing than entanglement can be realized in molecules. For nanostructures, we have been able to show that IT measures can be successfully employed to analyse the growing behaviour of PAMAM dendrimers supporting the dense-core model against the hollow-core one (Esquivel et al., 2009b, 2010g, 2011c).

In the Chapter we will present arguments based on the information content of L- and D-aminoacids to explain the biological preference toward homochirality. Besides, we present benchmark results for the information content of codons and aminoacids based on information-theoretical measures and statistical complexity factors which allow to elucidate the coding links between these building blocks and their selectivity.

2. Information-theoretical measures and complexities

In the independent-particle approximation, the total density distribution in a molecule is a sum of contribution from the electrons in each of the occupied orbitals. This is the case in both r-space and p-space, position and momentum respectively. In momentum space, the total electron density, (p), is obtained through the molecular momentals (momentum-space orbitals) $\phi_i(p)$, and similarly for the position-space density, $\rho(r)$, through the molecular position-space orbitals $\phi_i(r)$. The momentals can be obtained by three-dimensional Fourier transformation of the corresponding orbitals (and conversely)

$$\varphi_i(\mathbf{p}) = (2\pi)^{-3/2} \int d\mathbf{r} \exp(-i\mathbf{p} \cdot \mathbf{r}) \phi_i(\mathbf{r})$$ (1)

Standard procedures for the Fourier transformation of position space orbitals generated by ab-initio methods have been described (Rawlings & Davidson, 1985). The orbitals employed in ab-initio methods are linear combinations of atomic basis functions and since analytic expressions are known for the Fourier transforms of such basis functions (Kaijser & Smith, 1997), the transformation of the total molecular electronic wavefunction from position to momentum space is computationally straightforward (Kohout, 2007).

As we mentioned in the introduction, the LMC complexity is defined through the product of two relevant information-theoretic measures. So that, for a given probability density in position space, $\rho(r)$, the C(LMC) complexity is given by (Feldman & Crutchfield, 1998; Lamberti et al., 2004; Anteonodo & Plastino, 1996; Catalán et al., 2002; Martin et al., 2003):

$$C_r(LMC) = D_r e^{S_r} = D_r L_r$$ (2)

where Dr is the disequilibrium (Carbó-Dorca et al., 1980; Onicescu, 1966)

$$D_r = \int \rho^2(\mathbf{r})d\mathbf{r} \tag{3}$$

and S is the Shannon entropy (Shannon & Weaver, 1949)

$$S_r = -\int \rho(\mathbf{r})\ln\rho(\mathbf{r})d^3\mathbf{r} \tag{4}$$

from which the exponential entropy $L_r = e^{S_r}$ is defined. Similar expressions for the *LMC* complexity measure in the conjugated momentum space might be defined for a distribution $\gamma(p)$

It is important to mention that the *LMC* complexity of a system must comply with the following lower bound (López-Rosa et al., 2009):

$$C_p(LMC) = D_p e^{S_p} = D_p L_p \tag{5}$$

The *FS* complexity in position space, $Cr(FS)$, is defined in terms of the product of the Fisher information (Fisher, 1925; Frieden, 2004)

$$C(LMC) \geq 1 \tag{6}$$

and the power entropy (Angulo et al. 2008a; Sen et al., 2007a) in position space, Jr

$$I_r = \int \rho(\mathbf{r})|\vec{\nabla}\ln\rho(\mathbf{r})|^2 d^3\mathbf{r} \tag{7}$$

which depends on the Shannon entropy defined above. So that, the *FS* complexity in position space is given by

$$J_r = \frac{1}{2\pi e}e^{\frac{2}{3}S_r}, \tag{8}$$

and similarly

$$C_r(FS) = I_r \cdot J_r \tag{9}$$

in momentum space.

Let us remark that the factors in the power Shannon entropy J are chosen to preserve the invariance under scaling transformations, as well as the rigorous relationship (Dembo et al., 1991).

$$C_p(FS) = I_p \cdot J_p \tag{10}$$

with n being the space dimensionality, thus providing a universal lower bound to FS complexity. The definition in Eq. (8) corresponds to the particular case $n=3$, the exponent containing a factor $2/n$ for arbitrary dimensionality.

It is worthwhile noting that the aforementioned inequalities remain valid for distributions normalized to unity, which is the choice that it is employed throughout this work for the 3-dimensional molecular case.

Aside of the analysis of the position and momentum information measures, we have considered it useful to study these magnitudes in the product rp-space, characterized by the probability density $f(r, p) = \rho(r)\gamma(p)$, where the complexity measures are defined as

$$C(FS) \geq n \tag{11}$$

and

$$C_{rp}(LMC) = D_{rp}L_{rp} = C_r(LMC)C_p(LMC), \tag{12}$$

From the above two equations, it is clear that the features and patterns of both LMC and FS complexity measures in the product space will be determined by those of each conjugated space. However, the numerical analyses carried out in the next section, reveal that the the momentum space contribution plays a more relevant role as compared to the one in position space.

We have also evaluated some reactivity parameters that may be useful to analyze the chemical reactivity of the aminoacids. So that, we have computed several reactivity properties such as the ionization potential (IP), the hardness (η) and the electrophilicity index (ω). These properties were obtained at the Hartree-Fock level of theory (HF) in order to employ the Koopmans' theorem (Koopmans, 1933; Janak, 1978), for relating the first vertical ionization energy and the electron affinity to the HOMO and LUMO energies, which are necessary to calculate the conceptual DFT properties. Parr and Pearson, proposed a quantitative definition of hardness (η) within conceptual DFT (Parr & Yang, 1989):

$$C_{rp}(FS) = I_{rp}J_{rp} = C_r(FS)C_p(FS), \tag{13}$$

where ε denotes the frontier molecular orbital energies and S stands for the softness of the system. It is worth mentioning that the factor $1/2$ in Eq. (14) was put originally to make the hardness definition symmetrical with respect to the chemical potential (Parr & Pearson, 1983)

$$\eta = \frac{1}{2S} \approx \frac{\varepsilon_{LUMO} - \varepsilon_{HOMO}}{2} \tag{14}$$

although it has been recently disowned (Ayer et al. 2006: Pearson, 1995). In general terms, the chemical hardness and softness are good descriptors of chemical reactivity. The former has been employed (Ayer et al. 2006: Pearson, 1995; Geerlings et al., 2003) as a measure of the reactivity of a molecule in the sense of the resistance to changes in the electron distribution of the system, i.e., molecules with larger values of η are interpreted as being the least reactive ones. In contrast, the S index quantifies the polarizability of the molecule (Ghanty & Ghosh, 1993; Roy et al., 1994; Hati & Datta, 1994; Simon-Manso & Fuentealba, 1998) and hence soft molecules are more polarizable and possess predisposition to acquire additional electronic charge (Chattaraj et al., 2006). The chemical hardness η is a central quantity for use in the study of reactivity through the hard and soft acids and bases principle (Pearson, 1963; Pearson, 1973; Pearson, 1997).

The electrophilicity index (Parr et al., 1999), ω, allows a quantitative classification of the global electrophilic nature of a molecule within a relative scale. Electrophilicity index of a system in terms of its chemical potential and hardness is given by the expression

$$\mu = \left(\frac{\partial E}{\partial N} \right)_{v(r)} = \frac{\varepsilon_{LUMO} + \varepsilon_{HOMO}}{2} \tag{15}$$

The electrophilicity is also a good descriptor of chemical reactivity, which quantifies the global electrophilic power of the molecules -predisposition to acquire an additional electronic charge- (Parr & Yang, 1989).

3. Aminoacids

The exact origin of homochirality is one of the great unanswered questions in evolutionary science; such that, the homochirality in molecules has remained as a mystery for many years ago, since Pasteur. Any biological system is mostly composed of homochiral molecules; therefore, the most well-known examples of homochirality is the fact that natural proteins are composed of L-amino acids, whereas nucleic acids (RNA or DNA) are composed of D-sugars (Root-Bernstein, 2007; Werner, 2009; Viedma et al., 2008). The reason for this behavior continues to be a mystery. Until today not satisfactory explanations have been provided regarding the origin of the homochirality of biological systems; since, the homochirality of the amino acids is critical to their function in the proteins. If proteins (with L-aminoacids) had a non-homochiral behavior (with few D-enantiomers in random positions) they would not present biological functionality It is interesting to mention that L-aminoacids can be synthesized by use of specific enzymes, however, in prebiotic life these processes remain unknown. The same problem exists for sugars which have the D configuration. (Hein and Blackmond, 2011; Zehnacker et al., 2008; Nanda and DeGrado, 2004).

On the other hand, the natural amino acids contain one or more asymmetric carbon atoms, except the glycine. Therefore, the molecules are two nonsuperposable mirror images of each other; i.e., representing right-handed (D enantiomer) and left-handed (L enantiomer) structures. It is considered that the equal amounts of D- and L- amino acids existed on primal earth before the emergence of life. Although the chemical and physical properties of L-and D amino acids are extremely similar except for their optical character, the reason of the exclusion of D-amino acids and why all living organisms are now composed predominantly of L-amino acids are not well-known: however, the homochirality is essential for the development and maintenance of life (Breslow, 2011; Fujii et al., 2010; Tamura, 2008). The essential property of α-aminoacids is to form linear polymers capable of folding into 3-dimensional structures, which form catalytic active sites that are essential for life. In the process, aminoacids behave as hetero bifunctional molecules, forming polymers via head to tail linkage. In contrast, industrial nylons are often prepared from pairs of homo-bifunctional molecules (such as diamines and dicarboxylic acids), the use of a single molecule containing both linkable functionalities is somewhat simpler (Cleaves, 2010; Weber and Miller, 1981; Hicks, 2002).

The concept of chirality in chemistry is of paramount interest because living systems are formed of chiral molecules of biochemistry is chiral (Proteins, DNA, amino acids, sugars and many natural products such as steroids, hormones, and pheromones possess chirality). Indeed, amino acids are largely found to be homochiral (Stryer, 1995) in the L form. On the other hand, most biological receptors and membranes are chiral, many drugs, herbicides, pesticides and other biological agents must themselves possess chirality. Synthetic processes ordinarily produce a 50:50 (racemic) mixture of left-handed and right-handed molecules (so-called enantiomers), and often the two enantiomers behave differently in a biological system.

On the other hand, a major topic of research has been to study the origin of homochirality. In this respect, biomembranes have played an important role for the homochiraility of biopolymers. One of the most intriguing problems in life sciences is the mechanism of symmetry breaking. Many theories have been proposed on these topics and in the attempt to explain the amplification of a first enantiomeric imbalance to the enantiopurity of biomolecules (Bombelli et al., 2004). In all theories on symmetry breaking and on enantiomeric excess amplification little attention has been paid to the possible role of biomembranes, or of simple self-aggregated systems that may have acted as primitive biomembranes. Nevertheless, it is possible that amphiphilic boundary systems, which are considered by many scientists as intimately connected to the emergence and the development of life (Avalos et al. 2000; Bachmann et al., 1992), had played a role in the history of homochirality in virtue of recognition and compartmentalization phenomena (Menger and Angelova, 1998). In general, the major reason for the different recognition of two enantiomers by biological cells is the homochirality of biomolecules such as L-amino acids and D-sugars. The diastereomeric interaction between the enantiomers of a bioactive compound and the receptor formed from a chiral protein can cause different physiological responses. The production technology of enantiomerically enriched bioactive compounds one of the most important topics in chemistry. There is great interest in how and when biomolecules achieved high enantioenrichment, including the origin of chirality from the standpoint of chiral chemistry (Zehnacker et al., 2008; Breslow, 2011; Fujii et al., 2010; Tamura, 2008; Arnett and Thompson, 1981)

3.1. Physical and information-theoretical properties

Figure l illustrates a Venn diagram (Livingstone & Barton, 1993; Betts & Russell, 2003) which is contained within a boundary that symbolizes the universal set of 20 common amino acids (in one letter code). The amino acids that possess the dominant properties—hydrophobic, polar and small (< 60 Å3)—are defined by their set boundaries. Subsets contain amino acids with the properties aliphatic (branched sidechain non-polar), aromatic, charged, positive, negative and tiny (<35 Å 3). Shaded areas define sets of properties possessed by none of the common amino acids. For instance, cysteine occurs at two different positions in the Venn diagram. When participating in a disulphide bridge (C_{S-S}), cysteine exhibits the properties 'hydrophobic' and 'small'. In addition to these properties, the reduced form (C_{S-H}) shows polar character and fits the criteria for membership of the 'tiny' set. Hence, the Venn diagram (Figure l) assigns multiple properties to each amino acid; thus lysine has the property hydrophobic by virtue of its long sidechain as well as the properties polar, positive and charged. Alternative property tables may also be defined. For example, the amino acids might simply be grouped into non-intersecting sets labelled, hydrophobic, charged and neutral.

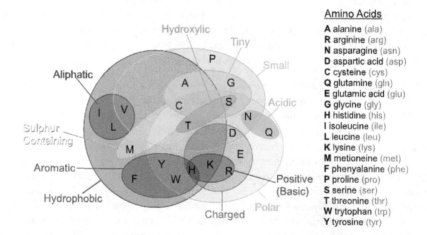

Figure 1. Venn diagram (Livingstone & Barton, 1993; Betts & Russell, 2003) of boundaries that symbolizes the universal set of 20 common amino acids (b). The Venn diagram in Fig. (1a) may be simply encoded as the property table or index shown here where the rows define properties and the columns refer to each amino acid.

In order to perform a theoretical-information analysis of L- and D-aminoacids we have employed the corresponfing L-enantiomers reported in the Protein Data Bank (PDB), which provide a standard representation for macromolecular structure data derived from X-ray diffraction and NMR studies. In a second stage, the D-type enantiomers were obtained from the L-aminoacids by interchanging the corresponding functional groups (carboxyl and amino) of the α-carbon so as to represent the D-configuration of the chiral center, provided that steric impediments are taken into account. The latter is achieved by employing the Ramachandran (Ramachandran et al, 1963) map, which represent the phi-psi torsion angles for all residues in the aminoacid structure to avoid the steric hindrance. Hence, the backbone of all of the studied aminoacids represent possible biological structures within the allowed regions of the Ramachandran. In the third stage, an electronic structure optimization of the geometry was performed on all the enantiomers for the twenty essential aminoacids so as to obtain structures of minimum energy which preserve the backbone (see above). In the last stage, all of the information-theoretic measures were calculated by use of a suite of programs which have been discussed elsewhere (Esquivel et al., 2012).

In Figures 2 through 4 we have depicted some selected information-theoretical measures and complexities in position space versus the number of electrons and the energy. For instance, it might be observed from Fig. 2 that the Shannon entropy increases with the number of electrons so that interesting properties can be observed, e.g., the aromatic ones possess more delocalized densities as the rest of the aminoacids (see Figure 1B) which confer specific chemical properties. On the other hand, the disequilibrium diminishes as the number of electron increases (see Fig. 2), which can be related to the chemical stability of the aminoacids, e.g., cysteine and metionine show the larger values (see Fig. 2) which is in agreement with the biological evidence in that both molecules play mutiple functions in proteins, chemical as well as structural, conferring the higher reactivity that is recognized to both molecules. In contrast, aromatic aminoacids (see Fig 1B) are the least reactive, which is in agreement with the lower disequilibrium values that are observed form Fig 2.

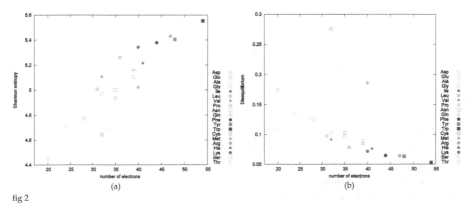

fig 2

Figure 2. Shannon entropy (left) and disequilibrum (right), both in position space, versus the number of electrons, for the set of 20 aminoacids.

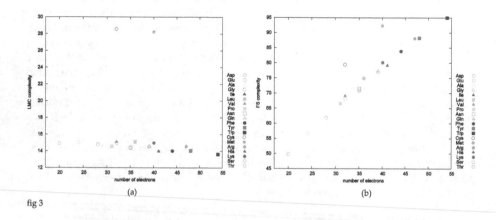

fig 3

Figure 3. LMC- (left) and FS-complexities (right), both in position space, versus the number of electrons, for the set of 20 aminoacids.

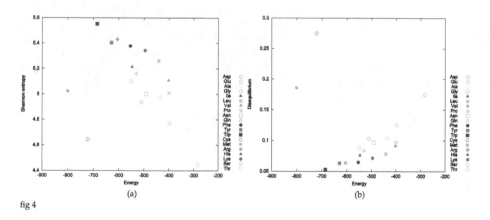

fig 4

Figure 4. LMC- (left) and FS-complexities (right), both in position space, versus the total energy, for the set of 20 aminoacids.

In Figures 3 we have plotted the *LMC* and *FS* complexities versus the number of electrons for the twenty aminoacids where we can observe that *LMC* complexity disntinguishes two different groups of aminoacids, where the more reactive (met and cys) hold larger values. In contrast, *FS* complexity behaves linearly with the number of electrons where the aromatic aminoacids possess the larger values and hence represent the more complex ones. Furthemore, the behavior of the *LMC* and *FS* complexities with respect to the total energy is analyzed in Figures 4, to note that *LMC* complexity characterizes two different groups of aminoacids where the most reactive (cys and met) possess the largest values, which incidentally hold the largest energies (negatively). A different behavior is observed for the *FS* complexity in that the smaller values correspond to the less energetic aminoacids. It is worthy to mention that the *FS* complexity is related to the Fisher information measure (Eq. 7) which depends on the local behavior of the position space density, i..e., simpler molecules present more ordered chemical structures, and hence these kind of aminoacids are expected to be less complex, e.g., the small and the tiny ones (Ser, Ala, Thr).

3.2. Homochirality

In Figures 5 through 8 we have analyzed the homochiral behavior of all aminoacids by plotting the difference between the L and the D values of several physical properties (energy, ionization potential, hardness, electrophilicity) and some relevant information-theoretical measures (Shannon entropy, Fisher, LMC- and FS-complexity). From Figures 5 and 6 one can readily observe that none of the physical properties studied in this work show a uniform enantiomeric behavior, i.e., it is not possible to distinguish the L-aminoacids from the D-ones by using an specific physical property. In contrast, the L-aminoacids can be uniquely characterize d from the D-ones when informatic-theoretical measures are employed (see Figures 7 and 8) and this is perhaps the most interesting result obtained from our work. To the best of our kowledge no similar observations have been reported elsewhere, showing strong evidence of the utility of Information Theory tools for decoding the essential blocks of life.

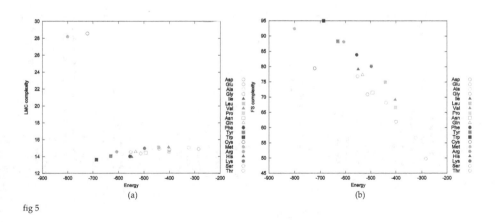

fig 5

Figure 5. Differences (L-property – D-property) for the energy (in a.u., left) and the ionization potential (in a.u., right) for the set of 20 aminoacids.

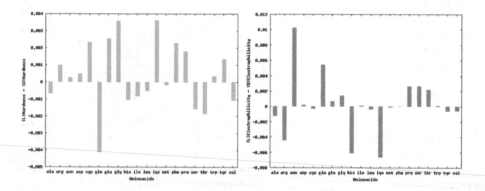

Figure 6. Differences (L-property – D-property) for the hardness (in a.u., left, Eqn. 14) and the electrophilicity (in a.u., right, Eqn. 16) for the set of 20 aminoacids.

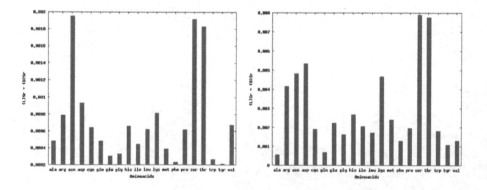

Figure 7. Differences (L-property – D-property) for the Shannon entropy (Eqn. 4, left) and the Fisher information (right, Eqn. 7), both in position space, for the set of 20 aminoacids.

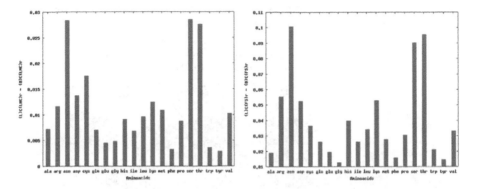

Figure 8. Differences (L-property – D-property) for the LMC (left, Eqn. 2) and the FS-complexities (right, Eqn. 9), both in position space, for the set of 20 aminoacids.

4. Genetic code

4.1. Codons

The genetic code refers to a nearly universal assignment of codons of nucleotides to amino acids. The codon to amino acid assignment is realized through: (i) the code adaptor molecules of transfer RNAs (tRNAs) with a codon's complementary replica (anticodon) and the corresponding amino acid attached to the 3' end, and (ii) aminoacyl tRNA synthetases (aaRSs), the enzymes that actually recognize and connect proper amino acid and tRNAs. The origin of the genetic code is an inherently difficult problem (Crick, 1976). Taking into a count that the events determining the genetic code took place long time ago, and due to the relative compactness of the present genetic code. The degeneracy of the genetic code implies that one or more similar tRNA can recognize the same codon on a messenger mRNA. The number of amino acids and codons is fixed to 20 amino acids and 64 codons (4 nucleotides, A.C.U.G per three of each codon) but the number of tRNA genes varies widely 29 to 126 even between closely related organisms. The frequency of synonymous codon use differs between organisms, within genomes, and along genes, a phenomenon known as CUB (codon usage bias) (Thiele et al., 2011).

Sequences of bases in the coding strand of DNA or in messenger RNA possess coded instructions for building protein chains out of amino acids. There are 20 amino acids used in making proteins, but only four different bases to be used to code for them. Obviously one base can't code for one amino acid. That would leave 16 amino acids with no codes. By taking two bases to code for each amino acid, that would still only give you 16 possible codes (TT, TC, TA, TG, CT, CC, CA and so on) – that is, still not enough. However, by taking three bases per amino acid, that gives you 64 codes (TTT, TTC, TTA, TTG, TCT, TCC and so on). That's enough to code for everything with lots to spare. You will find a full table of these below. A three base sequence in DNA or RNA is known as a codon.

		second base in codon			
		U	C	A	G	
		UUU Phe	UCU Ser	UAU Tyr	UGU Cys	U
	U	UUC Phe	UCC Ser	UAC Tyr	UGC Cys	C
		UUA Leu	UCA Ser	UAA stop	UGA stop	A
		UUG Leu	UCG Ser	UAG stop	UGG Trp	G
first base in codon	C	CUU Leu	CCU Pro	CAU His	CGU Arg	U
		CUC Leu	CCC Pro	CAC His	CGC Arg	C
		CUA Leu	CCA Pro	CAA Gln	CGA Arg	A
		CUG Leu	CCG Pro	CAG Gln	CGG Arg	G
	A	AUU Ile	ACU Thr	AAU Asn	AGU Ser	U
		AUC Ile	ACC Thr	AAC Asn	AGC Ser	C
		AUA Ile	ACA Thr	AAA Lys	AGA Arg	A
		AUG Met	ACG Thr	AAG Lys	AGG Arg	G
	G	GUU Val	GCU Ala	GAU Asp	GGU Gly	U
		GUC Val	GCC Ala	GAC Asp	GGC Gly	C
		GUA Val	GCA Ala	GAA Glu	GGA Gly	A
		GUG Val	GCG Ala	GAG Glu	GGG Gly	G

Table 1. Various combinations of three bases in the coding strand of DNA are used to code for individual amino acids - shown by their three letter abbreviation

The codes in the coding strand of DNA and in messenger RNA aren't, of course, identical, because in RNA the base uracil (U) is used instead of thymine (T). Table 1 shows how the various combinations of three bases in the coding strand of DNA are used to code for individual amino acids - shown by their three letter abbreviation. The table is arranged in such a way that it is easy to find any particular combination you want. It is fairly obvious how it works and, in any case, it doesn't take very long just to scan through the table to find what you want. The colours are to stress the fact that most of the amino acids have more than one code. Look, for example, at leucine in the first column. There are six different codons all of which will eventually produce a leucine (Leu) in the protein chain. There are also six for serine (Ser). In fact there are only two amino acids which have only one sequence of bases to code for them - methionine (Met) and tryptophan (Trp). Note that three codons don't have an amino acid but "stop" instead. For obvious reasons these are known as stop codons. The stop codons in the RNA table (UAA, UAG and UGA) serve as a signal that the end of the chain has been reached during protein synthesis. The codon that marks the start of a protein chain is AUG, that's the amino acid, methionine (Met). That ought to mean that every protein chain must start with methionine.

4.2. Physical and information-theoretical properties

An important goal of the present study is to characterize the biological units which codify aminoacids by means of information-theoretical properties. To accomplished the latter we have depicted in Figures 9 through 13 the Shannon entropy, Disequilibrium, Fisher and the LMC and FS complexities in position space as the number of electron increases, for the group of the 64 codons. A general observation is that all codons hold similar values for all these properties as judging for the small interval values of each graph. For instance, the Shannon entropy values for the aminoacids (see Figure 2) lie between 4.4 to 5.6, whereas the corresponding values for the codons (see Figure 9) lie between 6.66 to 6.82, therefore this information measure serves to characterize all these bilogical molecules, providing in this way the first benchmark informational results for the building blocks of life. Further, it is interesting to note from Figures 9 and 10 that entropy increases with the number of electrons (Fig. 9) whereas the opposite behavior is observed for the Disequilibrium measure. Besides, we may note from these Figures an interesting codification pattern within each isolelectronic group of codons where one may note that an exchange of one nucleotide seems to occur, e.g., as the entropy increases in the 440 electron group the following sequence is found: UUU to (UUC, UCU, CUU) to (UCC, CUC, CCU) to CCC. Similar observations can be obtained from Figures 10 and 11 for D and I, respectively. In particular, Fisher information deserves special analysis, see Figure 11, from which one may observe a more intricated behavior in which all codons seem to be linked across the plot, i.e., note that for each isoelectronic group codonds exchange only one nucleotide, e.g., in the 440 group codons change from UUU to (UUC, UCU, CUU) to (UCC, CUC, CCU) to CCC as the Fisher measure decreaes. Besides, as the Fisher measure and the number of electrons increase linearly a similar exchange is observed, eg., from AAA to (AAG, AGA, GAA) to (AGG, GAG, GGA) to GGG. We believe that the above observations deserve further studies since a codification pattern seems to be apparent.

In Figures 12 and 13 we have depicted the LMC and FS complexities, respectively, where we can note that as the number of electron increases the LMC complexity decreases and the opposite is observed for the FS complexity. It is worth mentioning that similar codification patternsm, as the ones above discussed, are observed for both complexities. Furthermore, we have found interesting to show similar plots in Figures 14 and 15 where the behavior of both complexities is shown with respect to the total energy. It is observed that as the energy increases (negatively) the LMC complexity decreases whereas the FS complexity increases. Note that similar codification patterns are observed in Figure 15 for the FS complexity.

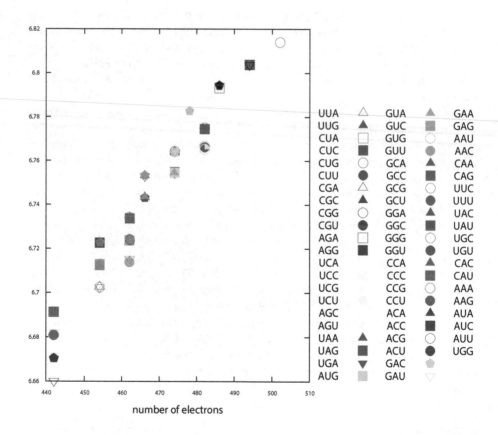

Figure 9. Shannon entropy values in position spaces as a function of the number of electrons for the set of 64 codons.

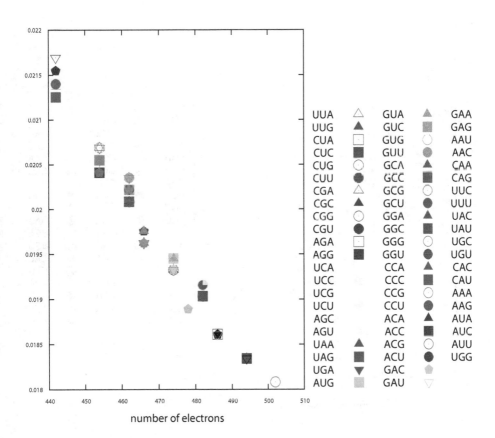

number of electrons

Figure 10. Disequilibrium in position spaces (Eqn. 3)as a function of the number of electrons for the set of 64 codons

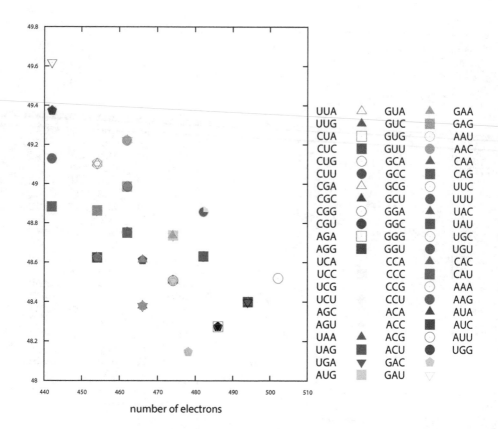

Figure 11. Fisher information in position spaces (Eqn. 7) as a function of the number of electrons for the set of 64 codons

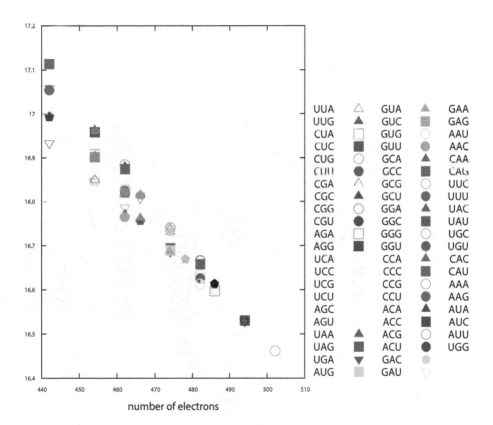

Figure 12. LMC-complexity in position spaces (Eqn.12) as a function of the number of electrons for the set of 64 co-dons

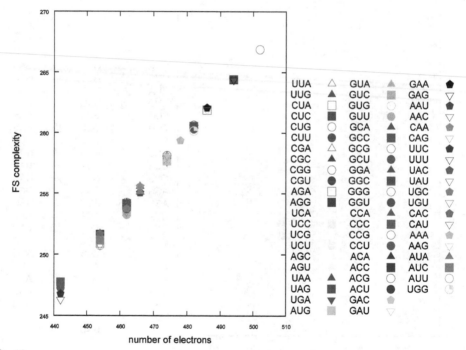

fig 13

Figure 13. FS-complexity in position spaces (Eqn. 7) as a function of the number of electrons for the set of 64 codons

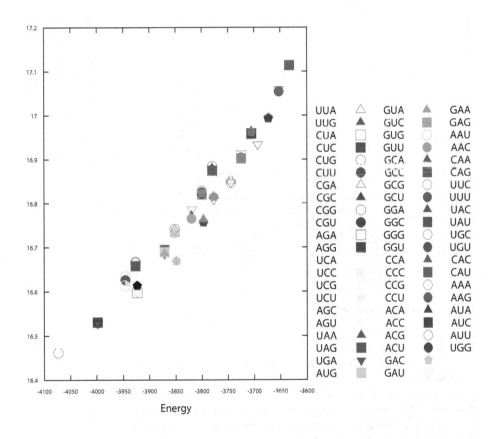

Figure 14. LMC-complexity in position spaces as a function of the total energy for the set of 64 codons

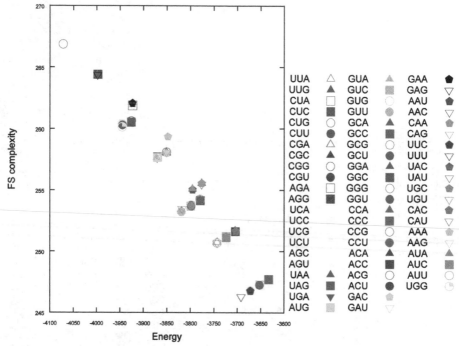

fig 15

Figure 15. FS-complexity in position space as a function of the total energy for the set of 64 codons.

5. Concluding remarks

We have shown throughout this Chapter that information-theoretical description of the fundamental biological pieces of the genetic code: aminoacids and codons, can be analysed in a simple fashion by employing Information Theory concepts such as local and global information measures and statistical complexity concepts. In particular, we have provided for the first time in the literature with benchmark information-theoretical values for the 20 essential aminoacids and the 64 codons for the nucleotide triplets. Throughout these studies, we believe that information science may conform a new scientific language to explain essential aspects of biological phenomena. These new aspects are not accessible through any other standard methodology in quantum chemistry, allowing to reveal intrincated mechanisms in which chemical phenomena occur. This envisions a new area of research that looks very promising as a standalone and robust science. The purpose of this research is to provide fertile soil to build this nascent scientific area of chemical and biological inquiry through information-theoretical concepts towards the science of the so called Quantum Information Biology.

Acknowledgements

We wish to thank José María Pérez-Jordá and Miroslav Kohout for kindly providing with their numerical codes. We acknowledge financial support through Mexican grants from CONACyT, PIFI, PROMEP-SEP and Spanish grants MICINN projects FIS2011-24540, FQM-4643 and P06-FQM-2445 of Junta de Andalucía. J.A., J.C.A., R.O.E. belong to the Andalusian researchs groups FQM-020 and J.S.D. to FQM-0207. R.O.E. wishes to acknowledge financial support from the CIE-2012. CSC., acknowledges financial support through PAPIIT-DGAPA, UNAM grant IN117311. Allocation of supercomputing time from Laboratorio de Supercómputo y Visualización at UAM, Sección de Supercomputacion at CSIRC Universidad de Granada, and Departamento de Supercómputo at DGSCA-UNAM is gratefully acknowledged.

Author details

Rodolfo O. Esquivel[1,6], Moyocoyani Molina-Espíritu[1], Frank Salas[1], Catalina Soriano[2], Carolina Barrientos[3], Jesús S. Dehesa[4,6] and José A. Dobado[5]

1 Departamento de Química, Universidad Autónoma Metropolitana-Iztapalapa, México D.F., México

2 Laboratorio de Química Computacional, FES-Zaragoza, Universidad Nacional Autónoma de México, C.P. 09230 Iztapalapa, México, D.F., México

3 Facultad de Bioanálisis-Veracruz, Universidad Veracruzana, Lab. de Química y Biología Experimental, Veracruz, México

4 Departamento de Física Atómica, Molecular y Nuclear, Universidad de Granada, Granada, Spain

5 Grupo de Modelización y Diseño Molecular, Departamento de Química Orgánica, Universidad de Granada, Granada, Spain

6 Instituto "Carlos I" de Física Teórica y Computacional, Universidad de Granada, Granada, Spain

References

[1] Angulo, J. C, Antolin, J, Sen, K. D, 2008a, Phys, , & Lett, A.

[2] Angulo, J. C. ; J, Antolín, , & Chem, J. Phys., 128, 164109 2008

[3]   Angulo, J. C, Antolin, J, & Esquivel, R. O. (2010d). Chapter in the monograph'Statistical Complexity: Applications in Electronic Structure', ed. K.D. Sen (Springer, United Kingdom, 2010) 978-9-04813-889-0

[4]   Anteonodo, C, & Plastino, A. (1996). Phys. Lett. A, 223, 348 1996.

[5]   Antolín, J, & Angulo, J. C. Quantum Chem., 109, 586 2009.

[6]   Arnett, E. M, & Thompson, O. J. Am. Chem. Soc. , 1981(103), 968-970.

[7]   Avalos, M, Babiano, R, Cintas, P, Jiménez, J. L, & Palacios, J. C. (2000). Tetrahedron: Asymmetry , 2000(11), 2845-874.

[8]   Ayers, P. W. ; R. G, Parr, R. G, & Pearson, J. Chem. Phys. 2006, 124, 194107

[9]   Bachmann, P. A, Luisi, P. L, & Lang, J. Nature , 1992(357), 57-59.

[10] Bennet, C. H. (1988). The Universal Turing Machine a Half Century, edited by R. Herhen, Oxford University Press, Oxford, , 227-257.

[11] Betts, M. J, & Russell, R. B. (2003). In Bioinformatics for Geneticists, M.R. Barnes, I.C. Gray eds, Wiley, 2003

[12] Bombelli, C, & Borocci, S. Lupi, Federica.; Mancini, G.; Mannina, L.; Segre, A. L.; Viel, Stéphane ((2004). J. Am. Chem. Soc. , 2004(126), 13354-13362.

[13] Borgoo, A, De Proft, F, Geerlings, P, & Sen, K. D. (2007). Chem. Phys. Lett. 444, 186 2007.

[14] Breslow, R. Thetrahedron Letters. (2011). , 2011(52), 4228-4232.

[15] Carbó-dorca, R. ; J, & Arnau, L. Leyda ((1980). Int. J. Quantum Chem., 17, 1185 1980.

[16] Carrera, E. M, Flores-gallegos, N, & Esquivel, R. O. Appl. Math. 233, 1483-1490 (2010).

[17] Catalán, R. G. ; J, & Garay, R. López-Ruiz ((2002). Phys. Rev. E, 66, 011102 2002.

[18] Chaitin, G, 1966, & Acm, J.

[19] Chattaraj, P. K ; U, & Sarkar, D. R. Roy, Chem. Rev., (2006).

[20] Chatzisavvas, K. Ch.; Ch.C. Moustakidis and C.P. Panos ((2005). J. Chem. Phys., 123, 174111 2005

[21] Cleaves, H. J. Biol. , 2010(263), 490-498.

[22] Crick, F, Brenner, S, Klug, A, & Pieczenik, G. (1976). Orig Life 1976, 7:389-397.

[23] Dembo, A, Cover, T. M, & Thomas, J. A. (1991). Information theoretic inequalities", IEEE Trans. Inform. Theory, , 37, 1501-1518.

[24] Dehesa, J. S. ; R, González-férez, P, Sánchez-moreno, J, & Phys, A.

[25]   Esquivel, R O, Flores-gallegos, N, Iuga, C, Carrera, E. M, Angulo, J. C, & Antolín, J. (2009a). Theoretical Chemistry Accounts 124, 445-460 (2009)

[26]   Esquivel, R O, & Flores-gallegos, N. Iuga, C; Carrera, E.M. ; Angulo, J.C. ; Antolín J. ((2010a). Physics Letters A , 374(2010), 948-951.

[27]   Esquivel, R O, Flores-gallegos, N, Sánchez-dehesa, J. S, Angulo, J. C, Antolín, J, Sen, K, 2010b, Phys,). J, & Chem, A. 2010(114), 1906-1916.

[28]   Esquivel, R. O, Liu, S, Angulo, J. C, Dehesa, J. S, Antolin, J, Molina-espíritu, M, 2011a, Phys,). J, Chem, A, Phys, J, & Chem, A.

[29]   Esquivel, R. O, Angulo, J. C, Antolin, J, Sánchez-dehesa, J, Flores-gallegos, N, & Lopez-rosa, S. Chem. Chem. Phys., 12, 7108-7116 (2010)

[30]   Esquivel, R. O, Molina-espíritu, M, Angulo, J. C, Antolín, J, Flores-gallegos, N, Dehesa, J. S, & Mol, b. Phys., (2011). , 109

[31]   Esquivel, R. O, Flores-gallegos, N, Carrera, E. M, Dehesa, J. S, Angulo, J. C, Antolín, J, & Soriano-correa, C. Sim., 35, (2009), 498-511;

[32]   Esquivel, R. O, Flores-gallegos, N, Carrera, E. M, & Soriano-correa, C. (2010g). Journal of Nano Research , 9, 1-15.

[33]   Esquivel, R. O, Flores-gallegos, N, Molina-espíritu, M, Plastino, A. R, Dehesa, J. S, Angulo, J. C, Antolín, J, Phys, c). J, & At, B. Mol. Opt. Phys. 44, 175101 ((2011).

[34]   Feldman, D. P. ; J. P. Crutchfield (1998)., Phys. Lett. A, 238, 244 (1998).

[35]   Fisher, R. A. (1925). Proc. Cambridge Philos. Soc., 22, 700 1925

[36]   Flores-gallegos, N, & Esquivel, R. O. Chem. Soc. (2008), 52(1), 19-30

[37]   Frieden, B. R. (2004). Science from Fisher Information (Cambridge University Press, Cambridge, 2004.

[38]   Fujii, N, Kaji, Y, Fujii, N, Nakamura, T, & Motoie, R. Mori, Yuhei.; Kinouchi, T ((2010). Chemistry & Biodiversity, , 2010(7), 1389-1397.

[39]   Geerlings, P. ; F, & De Proft, W. Langenaeker ((2003). Chem. Rev. 2003, 103, 1793.

[40]   Ghanty, T. K, & Ghosh, S. K. Chem., 1993, 97, 4951

[41]   Hati, S, Datta, D, & Phys, J. Chem., (1994).

[42]   He, C. (2010). Nat Chem Biol 2010; , 6, 863-865.

[43]   Hein, J. E, & Blackmond, D. G. (2011). Acc. Chem. Res. 201110.1021/ar200316n

[44]   Hicks, J. M. (2002). bk- ch001., 2002-0810.

[45]   Hornyak, I, & Nagy, A. (2007). Chem. Phys. Lett., 437, 132 2007.

[46]   Janak, J. F. (1978). Phys. Rev. B, 1978, 18, 7165.

[47] Kaijser, P. ; V. H. Smith, Jr. ((1997). Adv. Quant. Chem. 10, 37 1997.

[48] Kohout, M. (2007). program DGRID, 2007, version 4.2.

[49] Koopmans, T. A. (1933). Physica, 1933, 1, 104

[50] Kolmogorov, A. N. Probl. Inf. Transm. ((1965).

[51] Lamberti, P. W. ; M. T, Martin, A, & Plastino, O. A. Rosso ((2004). Physica A, 334, 119 2004.

[52] Livingstone, C D, & Barton, G J. (1993). CABIOS, 9 745 1993.

[53] López-rosa, S, Angulo, J. C, Antolín, J, 2009, & Physica, A.

[54] López-rosa, S, & Esquivel, R. O. Angulo, J.C ; Antolin, J.; Dehesa, J.S. ; Flores-Galle-gos, N. ((2010c). J. Chem. Theory Comput. , 6, 145-154.

[55] López-ruiz, R. ; H. L, & Mancini, X. Calbet ((1995). Phys. Lett. A, 1995, 209, 321.

[56] López-ruiz, R. Chem., 115, 215 2005.

[57] Lloyd, S, & Pagels, H. (1988). Ann. Phys. (N.Y.), 188, 186 1988.

[58] Martin, M. T, Plastino, A, & Rosso, O. A. (2003). Phys. Lett. A, 311, 126 2003.

[59] Menger, F. M, & Angelova, M. I. (1998). Acc. Chem. Res. , 1998(31), 789-797.

[60] Nagy, A. Phys., 2003, 119, 9401.

[61] Nagy, A, & Sen, K. D. (2006). Phys. Lett. A, 360, 291 2006.

[62] Nalewajski, R. (2003). Chem. Phys. Lett., 372, 28 2003.

[63] Nanda, V, & Degrado, W. F. Chem. Soc. , 2004(126), 14459-14467.

[64] Onicescu, O. Sci. Paris A 263, 25 1966.

[65] Parr, R. G, & Pearson, R. G. Chem. Soc., 1983, 105, 7512.

[66] Parr, R. G. ; W. Yang, ((1989). Density-Functional Theory of Atoms and Molecules (Oxford University Press: New York, 1989).

[67] Parr, R. G. ; L. V, Szentpály, S, & Liu, J. Am. Chem. Soc., 1999, 121, 1922.

[68] Pearson, R. G. Chem. Soc., 1963, 85, 3533

[69] Pearson, R. G. (1973). Hard and Soft Acids and Bases (Dowen, Hutchinson and Ross: Stroudsberg, 1973)

[70] Pearson, R. G. (1995). Inorg. Chim. Acta 1995, 240, 93.

[71] Pearson, R. G. (1997). Chemical Hardness; (Wiley-VCH; New York, 1997). Ramachan-dran GN, Ramakrishnan C, Sasisekharan J Mol Biol., 7:95-99., 1963

[72] Rawlings, D. C. ; E. R, & Davidson, J. Phys. Chem, 89, 969 1985.

[73] Romera, E. ; J. S, & Dehesa, J. Chem. Phys., 120, 8906 2004.

[74] Romera, E. ; P, Sánchez-moreno, J. S, & Dehesa, J. Math. Phys., 47, 103504 2006

[75] Rosso, O. A, Martin, M. T, Plastino, A, 2003, & Physica, A.

[76] Root-bernstein, R. (2007). BioEssays, , 2007(29), 689-698.

[77] Roy, R. ; A. K, Chandra, S, & Pal, J. Phys. Chem., 1994, 98, 10447

[78] Sen, K. D, Antolín, J, Angulo, J. C, 2007a, Phys, , & Rev, A.

[79] Sen, K. D. ; C. P, & Panos, K. Ch. Chatzisavvas and Ch. Moustakidis ((2007b). Phys. Lett. A, 364, 286 2007.

[80] Shalizi, C. R. ; K. L, & Shalizi, R. Haslinger, ((2004). Phys. Rev. Lett., 93, 118701 2004.

[81] Simon-manso, Y, Fuenteaelba, P, 1998, Phys, J, & Chem, A.

[82] Shannon, C. E, & Weaver, W. (1949). The Mathematical Theory of Communication (University of Illinois Press, Urbana, 1949.

[83] Stryer, L. (1995). Biochemistry, 4th ed., W. H. Freeman and Co.: New York, 1995.

[84] Tamura, K. (2008). BioSystems. , 2008(92), 91-98.

[85] Thiele, I, Fleming, R, Que, R, Bordbar, A, & Palsson, B. (2011). Nature Precedings 2011;

[86] Viedma, C, Ortiz, J. E, Torrea, T, Izumi, T, & Blackmond, D. G. Chem. Soc. , 2008(130), 15274-15275.

[87] Vitanyi, P. M. B, & Li, M. (2000). IEEE Trans. Inf. Theory, 46, 446 2000.

[88] Weber, A. L, & Miller, S. L. Evol. , 1981(17), 273-284.

[89] Werner, F. (2009). Colloids and Surfaces B: Biointerfaces, , 2009(74), 498-503.

[90] Yamano, T, 1995, & Physica, A.

[91] Yamano, T. Phys., 45, 1974 2004.

[92] Zehnacker, A, & Suhm, M. A. (2008). Angew. Chem. Int. Ed. , 2008(47), 6970-6992.

Chapter 7

A Novel Isospectral Deformation Chain in Supersymmetric Quantum Mechanics

Bjørn Jensen

Additional information is available at the end of the chapter

1. Introduction

Supersymmetric quantum mechanics (SUSYQM) has turned out to be surprisingly fertile field which is also able to successfully address challenges in traditional quantum mechanics and beyond. It has its roots in the works of Schrödinger, Infeld and Hull [1] on factorization methods of the Schrödinger equation. The term *supersymmetric* is due to a work by Witten [2] which brought these methods in contact with contemporary ideas in high energy physics. He showed in particular that the a factorized one-dimensional Schrödinger equation can accompany a super-Lie algebra thus providing a rich toy-model where features and concepts in supersymmetric quantum field theories can be studied in a greatly simplified context. A key ingredient in supersymmetric theories is that every bosonic state has a fermionic superpartner with all properties equal except the spin. In SUSYQM these states emerge as bosonic doublet states. The bosonic and the fermionic states are described in terms of the Schrödinger equation, but they interact with different physical potentials. These potentials are called partner potentials. Not completely surprising, knowing in advance the energy-eigenvalues and functions of the bosonic (fermionic) states the theory provides a map to the fermionic (bosonic) states with exactly the same energy-eigenvalues. Of key interest to us is that the physical partner potentials are expressed in terms of the same superpotential. These expressions are in general not unique. Different superpotentials can give rise to a particular physical potential in the fermionic (bosonic) sector. This does not imply that when the superpotential is changed (deformed) in such a way that the physical potential in the fermionic (bosonic) sector stays unchanged that the physical potential stays invariant in the bosonic (fermionic) sector. Whenever we deform a superpotential in the fermionic (bosonic) sector such that the fermionic (bosonic) potential is invariant the bosonic (fermionic) potential will generally change, but the theory nevertheless assures that the energy-eigenvalues in the bosonic (fermionic) sector stays the same. Such deformations are called isospectral deformations. They are the subject of this chapter.

Isospectral transformations in the context of SUSYQM has a long history exhibiting methods dating all the way back to Darboux [3]. The dominating approach is to study isospectral Hamilton operators. Different operator methods exist, but the main ones was brought under a single unifying principle by Pursey [4] with the use of isometric operators. A second approach to the study of isospectral transformations is what has been called deformation theory (see [5], e.g.). This is a more direct approach compared with the operator approach in that one studies deformations of the superpotential as described briefly above. It is rather surprising to note that this second approach has not been given much attention in the literature. To the knowledge of this author only one of the simplest deformations possible has been discussed to some extend. In a previous work [6] we initiated a work with the aim to remedy this situation. In [6] we showed that the isospectral deformation which has been considered in previous works is part of a more general deformation scheme. In this work we extend our results in [6]. We explicitly construct an in principle infinite recursively defined isospectral deformation chain where the deformation scheme in [6] emerges as the root of the chain.

This chapter is organized as follows. In the next section we very briefly review some of the basics of SUSYQM, mainly in order to fix notation. We define the notions of partner potentials, superpotential, isospectrality and supersymmetry. In section 2 we briefly remain ourselves about the results in [6]. In section 3 we define the recursive deformation scheme. We also discuss other various deformation schemes but show that a number of other canonical deformation schemes defined along the lines of our recursive scheme either do not allow a recursive structure or either reduces to our scheme. We apply our apparently rather unique recursive deformation scheme to the Coulomb potential and calculate several novel potentials. We summarize our findings and conclude in the last section. No attempt has been made to give an in depth review of the relevant literature due to its immense size. The works which have been acknowledged in the list of references have been so only because of their utility to this author.

2. SUSYQM - A very brief introduction

SUSYQM can in its most basic formulation be thought of as the following two factorizations of the Hamiltonian in the stationary Schrödinger equation in appropriate units

$$\begin{cases} (-\partial + W(x))(\partial + W(x)) \equiv A^+ A^- \equiv H_-, \\ (\partial + W(x))(-\partial + W(x)) \equiv A^- A^+ \equiv H_+. \end{cases} \tag{1}$$

Here ∂ is short hand for differentiation with respect to the single spatial coordinate x, and $W(x)$ is the so called superpotential. Both of these factorizations give rise to a Schrödinger equation, but with different potentials $V_-(x)$ and $V_+(x)$ (the so called superpotentials) given by

$$V_\pm(x) = W^2(x) \pm \partial W(x). \tag{2}$$

Let us denote the energy-eigenvalues and eigenstates associated with $H_\pm$ by $E_n^\pm$ and $\psi_n^\pm(x)$, respectively. Let $n = 0$ denote the ground state. We note that

$$A^\pm \psi_0^\pm(x) = 0 \Rightarrow H_\pm \psi_0^\pm(x) = 0. \tag{3}$$

The ground state eigenfunction is thus simply given by

$$\psi_0^{\pm}(x) \sim e^{\pm \int^x W(x)dx} \sim \frac{1}{\psi_0^{\mp}(x)} . \tag{4}$$

The factorization in Eq.(1) carries a symmetry which is not manifestly present in the usual form of the Schrödinger equation. This symmetry is made manifest when Eq.(1) is brought to a matrix form. Defining

$$Q^- \equiv \begin{pmatrix} 0 & 0 \\ A^- & 0 \end{pmatrix}, Q^+ \equiv \begin{pmatrix} 0 & A^+ \\ 0 & 0 \end{pmatrix} \tag{5}$$

we find that we naturally can construct a matrix-valued Hamiltonian H given by

$$H \equiv \begin{pmatrix} H_- & 0 \\ 0 & H_+ \end{pmatrix} = \begin{pmatrix} A^+ A^- & 0 \\ 0 & A^- A^+ \end{pmatrix} . \tag{6}$$

It is straightforward to verify that

$$H = Q^- Q^+ + Q^+ Q^- \equiv \{Q^-, Q^+\}, [Q^{\pm}, H] = 0, (Q^{\pm})^2 = 0. \tag{7}$$

This constitutes what is called a super-Lie algebra in contrast to an ordinary Lie algebra which only contains commutators. The commutator in Eq.(7) shows that $Q^{\pm}$ are generators of a symmetry which is left intact under time-translations generated by H. We call this symmetry the supersymmetry of the system.

The matrices above naturally act on a two-dimensional vector space with the natural representation

$$\begin{pmatrix} \psi_n^-(x) \\ \psi_{n-1}^+(x) \end{pmatrix} . \tag{8}$$

It is clear that

$$Q^- \begin{pmatrix} \psi_n^-(x) \\ 0 \end{pmatrix} = \begin{pmatrix} 0 \\ \psi_{n-1}^+(x) \end{pmatrix} , \tag{9}$$

$$Q^+ \begin{pmatrix} 0 \\ \psi_n^+(x) \end{pmatrix} = \begin{pmatrix} \psi_{n+1}^-(x) \\ 0 \end{pmatrix} . \tag{10}$$

Hence, $Q^{\pm}$ relate states with the same eigenvalue of H; the energy states are in other words degenerate. An orthogonal basis can naturally be taken to be states on the form

$$\begin{pmatrix} \alpha(x) \\ 0 \end{pmatrix}, \begin{pmatrix} 0 \\ \beta(x) \end{pmatrix} . \tag{11}$$

It is customary, due to the intimate relation to supersymmetric quantum field theory, to say that the first vector belongs to the bosonic sector and the other to the fermionic sector, even though no fermions appear in this theory. That $Q^\pm$ relate states corresponding to the same energy eigenvalue of the H-operator can also be seen on the level of the $H_\pm$ operators by noting that

$$H_+(A^-\psi_n^-) = E_n^- (A^-\psi_n^-). \tag{12}$$

Hence, given an eigenstate ψ_n^- of H_- with energy eigenvalue E_n^-, the state $A^-\psi_n^-$ is an eigenstate of H_+ with energy eigenvalue E_n^-. There is thus a one-to-one correspondence between bosonic and fermionic states with the same energy eigenvalue (Eq.(9-10) above). We call this property the isospectrality of SUSYQM. Much more can be said about SUSYQM, such as the role played by the vacuum in connection with isospectrality. However, for the purpose of this chapter this very brief exposition of some of the basics of SUSYQM is sufficient to fix notation and certain concepts.

3. A novel isospectral deformation chain

In [6] we introduced within the framework of SUSYQM an isospectral deformation on the form

$$W(x) \rightarrow \hat{W}_0(x) = F_0(x)W(x), \tag{13}$$

where $W(x)$ is some known superpotential and $F_0(x)$ some function to be determined by the isospectrality condition

$$\hat{V}_+(x) \equiv \hat{W}_0^2(x) + \hat{W}_0'(x) = W^2(x) + W'(x) \equiv V_+(x). \tag{14}$$

It was shown that Eq.(13) includes the only previously explored deformation of this kind, which has the form [5]

$$W(x) \rightarrow \hat{W}_0(x) = W(x) + f(x). \tag{15}$$

$f(x)$ is some function which is determined by Eq.(14). In this work we expand the deformation Eq.(13) in various directions and study the implications drawn from the isospectrality condition. We show in particular that the deformation Eq.(13) is the root of an infinitely long and recursively generated chain of deformations. Let us next briefly review some of the findings in [6].

3.1. Base deformations

The deformation Eq.(1) implies the following differential equation for $F_0(x)$ [6][1]

$$\frac{d}{dx}F_0(x) + (\frac{d}{dx}\ln W(x))F_0(x) + W(x)F_0^2(x) = W(x) + \frac{d}{dx}\ln W(x). \tag{16}$$

---

[1] We will often rewrite fractions on the form $W'(x)/W(x)$ as the logarithmic derivative of $W(x)$ as a formal tool. Caution must of course be exercised when using the corresponding expressions in actual computations.

This is the generalized Riccati equation [7] . If one particular solution $F_{00}(x)$ of Eq.(16) is known another solution is given by [8]

$$F_0(x) = F_{00}(x) + \frac{1}{X_0(x)},$$ (17)

where $X_0(x)$ solves the equation

$$\frac{d}{dx}X_0(x) - (\frac{d}{dx}\ln W(x) + 2F_{00}(x)W(x))X_0(x) = W(x).$$ (18)

Eq.(18) can be solved by elementary means. The resulting superpotential $\hat{W}_0(x)$ is given by [6]

$$\hat{W}_0(x) = (F_{00}(x) + \frac{1}{X_0(x)})W(x) \equiv \hat{W}_{00}(x) + \frac{1}{X_0(x)}W(x) =$$

$$= F_{00}(x)W(x) + \frac{e^{-2\int^x F_{00}(t)W(t)dt}}{C_{01} + \int^x e^{-2\int^u F_{00}(t)W(t)dt}du}.$$ (19)

C_{01} is an integration constant, which we will assume to be real. We have explicitly introduced upper integration limits in Eq.(19) in order to avoid sign ambiguities. This explains the difference in the sign in the denominator in Eq.(19) compared with Eq.(2.5) in [6] where the reverse order of integration in one of the integrals was implicitly assumed. We do not specify the lower integration limits in Eq.(19). These are not important, of course, since the values of the integrals there can essentially be absorbed into C_{01}. We can by simple inspection see that the particular solution $F_0(x) = 1$, the *identity deformation*, solves Eq.(16). With $F_{00}(x) = 1$ we identically rederive Eq.(15) and the corresponding expression discussed in [5]. The identity deformation corresponds to the limit $C_{01} \to \infty$ in Eq.(19) with $F_{00}(x) = 1$. In the limit $C_{01} \to \infty$ we generally get $\hat{W}_0(x) = \hat{W}_{00}(x)$. This deformation will play a pivotal role in this work; it will represent the base of a recursive scheme for generating novel isospectral deformations. We will therefore refer to a particular $\hat{W}_{00}(x)$ as *a base deformation* in the following.

In order to expand the space of concrete isospectral deformations further we transform Eq.(16) into an ordinary second order differential equation by the substitution

$$F_0(x) = \frac{1}{W(x)}\frac{d}{dx}\ln U_0(x).$$ (20)

This substitution gives rise to the following linear homogeneous second order differential equation

$$-\frac{d^2}{dx^2}U_0(x) + V_+(x)U_0(x) = 0.$$ (21)

This equation coincides of course with the zero-energy eigenfunction equation. However, keep in mind that U_0 is *not* in general to be identified with the eigenfunction of the system. This is of particular importance to remember in light of Eq.(16). The special solution $F_0(x) = 1$ is generated by the solution

$$U_0(x) \sim e^{\int^x W(t)dt} \, . \tag{22}$$

The particular solutions for $F_0(x)$ stemming from Eq.(21) can be fed into Eq.(19) (as $F_{00}(x)$) and thus expand the space of available concrete deformations. The physical potential $\hat{V}_-(x)$ generated by $\hat{W}_0(x)$ can in general thus be written [6][2]

$$\hat{V}_-(x) \equiv \hat{W}_0^2(x) - \hat{W}_0'(x) = \hat{W}_{00}^2(x) - \hat{W}_{00}'(x) +$$
$$+ \frac{4\hat{W}_{00}(x)e^{-2\int^x \hat{W}_{00}(t)dt}}{C_{01} + \int^x e^{-2\int^u \hat{W}_{00}(t)dt}du} + 2\left[\frac{e^{-2\int^x \hat{W}_{00}(t)dt}}{C_{01} + \int^x e^{-2\int^u \hat{W}_{00}(t)dt}du}\right]^2 \tag{23}$$

with

$$\hat{W}_{00}(x) = \frac{d}{dx}\ln U_0(x) \, . \tag{24}$$

3.2. Recursive linear deformations

Although the Riccati equation can be transformed into an ordinary second order differential equation the non-linearity of the Riccati equation allows for a solution space which is larger than the one associated with linear differential equations of second order, as became evident in the previous section. It is therefore natural to ask whether the non-linearity of the Riccati equation implies even more isospectral deformations than the ones we already have deduced [6]. We will explore this question in this and the next section.

3.2.1. The sum

Let us entertain the following idea. Assume that we have derived a particular base deformation $\hat{W}_{00}(x)$ from an explicitly given superpotential $W(x)$. Then assume that we *add* another term $F_1(x)W(x)$ (possibly multiplied with a constant) to that deformation such that we in principle get a novel deformation on the form $\hat{W}(x) = F_{10}(x)W(x) + \hat{W}_{00}(x)$. After determining $F_{10}(x)$ from the isospectrality condition Eq.(14) add yet another term of this kind to the deformation. Let us assume that this process can be repeated indefinitely. Will terms added in this manner give rise to novel deformations? We will in the following show that they do. This represents a recursive deformation scheme.

Following the basic idea, after m iterations we thus have the general recursive linear (in $W(x)$) deformation

$$\hat{W}_{m0}(x) = (\sum_{i=0}^{m} \lambda_i F_{i0}(x))W(x) = \lambda_m F_{m0}(x)W(x) + \hat{W}_{(m-1)0}(x) \, , \quad \lambda_0 \equiv 1. \tag{25}$$

---

[2] Note that the corresponding expression in [6] ((2.14)) is misprinted.

The λ_i's are assumed to be independent real constants. Starting with a known superpotential m consecutive applications of the isospectrality condition yields the following set of equations

$$
\begin{cases}
F_{00}'(x) + [\ln W(x)]' F_{00}(x) + W(x) F_{00}^2(x) = W(x) + (\ln W(x))', \\
F_{10}'(x) + [(\ln W(x))' + 2F_{00}(x)W(x)]F_{10}(x) + \lambda_1 W(x) F_{10}^2(x) = 0, \\
F_{20}'(x) + [(\ln W(x))' + 2(F_{00}(x) + \lambda_1 F_{10}(x))W(x)]F_{20}(x) + \lambda_2 W(x) F_{20}^2(x) = 0, \\
\vdots \\
\vdots \\
F_{m0}'(x) + [(\ln W(x))' + 2\hat{W}_{(m-1)0}(x))]F_{m0}(x) + \lambda_m W(x) F_{m0}^2(x) = 0.
\end{cases}
\tag{26}
$$

The first equation in Eq.(26) coincides of course per definition with Eq.(16). Note that $F_{j0}(x) = 1$ only solves the first equation in Eq.(26). Let us consider an arbitrary iteration level n ($\neq 0$) and make the following substitution in Eq.(26)

$$
F_{n0}(x) = \frac{1}{W(x)}(\ln U_n(x))'.
\tag{27}
$$

The equation for $F_n(x)$ can then be written

$$
U_n''(x) + (\lambda_n - 1)\frac{[U_n'(x)]^2}{U_n(x)} + 2\hat{W}_{(n-1)0}(x)U_n'(x) = 0.
\tag{28}
$$

This equation corresponds to Eq.(21) in the case when $n = 0$. It reduces in general to an ordinary linear differential equation only when $\lambda_n = 1, \forall n \neq 0$. We will focus on this special case in this work.

The general solution of Eq.(28) for arbitrary $n \neq 0$, and with λ_n set to unity, can be found by elementary means, and we deduce that

$$
F_{n0}(x)W(x) = \frac{C_{n2}e^{-2\int^x \hat{W}_{(n-1)0}(t)dt}}{C_{n1} + C_{n2}\int^x e^{-2\int^u \hat{W}_{(n-1)0}(t)dt}du} =
$$
$$
= \frac{d}{dx}\ln(C_{n1} + C_{n2}\int^x e^{-2\int^u \hat{W}_{(n-1)0}(t)dt}du).
\tag{29}
$$

C_{n1} and C_{n2} are integration constants, which we assume to be real. We can reduce the number of integration constants to one at each iteration level, but we will stick to the habit of explicitly writing down the actual number of constants in order to make it easier to compare the various formulas we deduce, which stem from both second and first order differential equations. We also note that the structure of $F_{n0}(x)$ implies that previous deformations are not regenerated in general. Of course, this does not exclude this possibility to arise, as we will see in Section 5. Hence, m in Eq.(25) has in principle no natural upper bound. From Eq.(25) and Eq.(29) we get the following expression for the superpotential at iteration level m

$$\hat{W}_{m0}(x) = \hat{W}_{00}(x) + \sum_{j=1}^{m} \frac{d}{dx} \ln(C_{j1} + C_{j2} \int^x e^{-2\int^u \hat{W}_{(j-1)0}(t)dt} du) =$$

$$= \hat{W}_{00}(x) + \frac{d}{dx} \ln \prod_{j=1}^{m} (C_{j1} + C_{j2} \int^x e^{-2\int^u \hat{W}_{(j-1)0}(t)dt} du) \equiv$$

$$\equiv \hat{W}_{00}(x) + \frac{d}{dx} \ln P_m(x). \tag{30}$$

From Eq.(30) we deduce that

$$e^{-2\int^x \hat{W}_{(j-1)0}(t)dt} = P_{j-1}^{-2}(x) e^{-2\int^x \hat{W}_{00}(t)dt} \; ; \; P_0^2(x) \equiv 1, j \neq 0, \tag{31}$$

such that

$$\begin{cases} P_1(x) = C_{11} + C_{12} \int^x e^{-2\int^u \hat{W}_{00}(t)dt} du, \\ P_2(x) = P_1(x)(C_{21} + C_{22} \int^x P_1^{-2}(u) e^{-2\int^u \hat{W}_{00}(t)dt} du), \\ \vdots \\ \vdots \\ P_m(x) = P_{m-1}(x)(C_{m1} + C_{m2} \int^x P_{m-1}^{-2}(u) e^{-2\int^u \hat{W}_{00}(t)dt} du). \end{cases} \tag{32}$$

Hence,

$$P_n(x) = \prod_{j=1}^{n} (C_{j1} + C_{j2} \int^x P_{j-1}^{-2}(u) e^{-2\int^u \hat{W}_{00}(t)dt} du). \tag{33}$$

This last form of the $P_n(x)$ functions neatly exhibits how the base deformation $\hat{W}_{00}(x)$ generates the higher order deformations. Some of the details we have deduced so far are presented in Figure 1.

Make the following substitution at each iteration level in Eq.(25)

$$F_{n0}(x) \rightarrow F_{n0}(x) + \frac{1}{X_n(x)}. \tag{34}$$

This implies (with the λ_m's reinstated in Eq.(25)) a generalized form $\hat{W}_m(x)$ of the superpotential $\hat{W}_{m0}(x)$

$$\hat{W}_{m0}(x) = \sum_{i=0}^{m} \lambda_i F_{i0} W(x) \Rightarrow \hat{W}_m(x) = \hat{W}_{m0}(x) + \sum_{i=0}^{m} \frac{\lambda_i}{X_i(x)} W(x). \tag{35}$$

$$\begin{cases} W \xrightarrow{\times F_0 = F_{00}} \hat{W}_{00} \xrightarrow{+F_{10}W} \hat{W}_{10} = \hat{W}_{00} + (\ln P_1)' \xrightarrow{+F_{20}W} \cdots \xrightarrow{+F_{m0}W} \hat{W}_{m0} \\[2em] W \xrightarrow{\times F_0 = 1} \hat{W}_{00} = W \xrightarrow{+F_{10}W} \hat{W}_{10} = W + (\ln P_1)' \xrightarrow{+F_{20}W} \cdots \xrightarrow{+F_{m0}W} \hat{W}_{m0} \end{cases}$$

Figure 1. The upper line depicts the solvable deformation chain Eq.(25) to iteration level m. There is no upper bound on m. The $F_{j0}(x)$ functions are given in Eq.(29). The $P_j(x)$ functions are given in Eq.(30) and Eq.(32). They are functions of a base deformation $\hat{W}_{00}(x)$. A base deformation $\hat{W}_{00}(x)$ is generated by the zero-energy Schrödinger equation interacting with the partner potential $V_+(x)$, Eq.(21). The second line depicts the important special case when $F_0(x) = 1$. This particular solution can be derived as a special case of Eq.(19) with $X_0^{-1} = 0$, which can be achieved by $C_{01} \to \infty$, and $F_{00}(x)$ determined by Eq.(20) and the solution Eq.(22). $\hat{W}_{10}(x)$ then coincides with Eq.(19) (when $F_{00}(x) = 1$ and C_{01} is finite in Eq.(19)); Eq.(19) is thus in this particular case regenerated by the scheme at the next recursion level, i.e.

From Eq.(26) we find that $X_n(x)$ satisfies the equation

$$\frac{d}{dx} X_n(x) - \left(\frac{d}{dx} \ln W(x) + 2\hat{W}_{n0}(x)\right) X_n(x) = \lambda_m W(x). \tag{36}$$

This equation is a generalization of Eq.(18). The n'th deformation term Eq.(29) thus changes into

$$F_{n0}(x)W(x) \to F_{n0}(x)W(x) + \frac{d}{dx} \ln\left(C_{n3} + \lambda_n \int^x e^{-2\int^u \hat{W}_{n0}(t)dt} du\right). \tag{37}$$

C_{n3} are integrations constants, which we assume to be real. Eq.(37) implies that the more general expression for the superpotential in Eq.(35) can be written as

$$\hat{W}_m(x) = \hat{W}_{m0}(x) + \frac{d}{dx} \ln Q_m(x), \tag{38}$$

where

$$Q_m(x) \equiv \prod_{i=0}^{m} \left(C_{i3} + \lambda_i \int^x e^{-2\int^u \hat{W}_{i0}(t)dt} du\right). \tag{39}$$

$m = 0$ in Eq.(38) ($\lambda_0 \equiv 1$) reproduces Eq.(19). In the special case when $\lambda_m = 1, \forall m$ in Eq.(25) we get

$$\hat{W}_m(x) = \hat{W}_{00}(x) + \frac{d}{dx} \ln P_m(x) + \frac{d}{dx} \ln Q_m(x). \tag{40}$$

When we compare the expressions for $(P_m(x))'$ and $(Q_m(x))'$ we find that they differ by just the last term in $(Q_m(x))'$.

3.2.2. The product

What happens if we in Eq.(25) assume a product structure instead of a sum structure ? Let us assume that we have determined a base deformation. Let this be the seed superpotential for the deformation

$$\hat{W}_{00}(x) \rightarrow \hat{W}_{10}(x) = F_{10}(x)\hat{W}_{00}(x) = F_{10}(x)F_{00}(x)W(x), \tag{41}$$

where $F_{10}(x)$ is some function to be determined by the isospectrality condition. This product scheme can of course in principle be repeated an arbitrary number m times

$$\hat{W}_{m0}(x) = (\prod_{i=0}^{m} F_{i0}(x))W(x) = F_{m0}(x)\hat{W}_{(m-1)0}. \tag{42}$$

This structure gives rise to the following set of equations

$$\begin{cases} F'_{00}(x) + (\ln W(x))'F_{00}(x) + W(x)F^2_{00}(x) = W(x) + (\ln W(x))', \\ F'_{10}(x) + (\ln \hat{W}_{00}(x))'F_{10}(x) + \hat{W}_{00}(x)F^2_{10}(x) = \frac{1}{F_{00}(x)}(W + (\ln W(x))'), \\ \vdots \\ \vdots \\ F'_{m0}(x) + (\ln \hat{W}_{(m-1)0}(x))'F_{m0}(x) + \hat{W}_{(m-1)0}(x)F^2_{m0}(x) = \frac{W(x)}{\hat{W}_{(m-1)0}(x)}(W(x) + (\ln W(x))'). \end{cases}$$
$$\tag{43}$$

Clearly, each iteration level depends on all the previous ones, and at each level we are dealing with a non-homogenous non-linear differential equation. Interestingly, by making the following substitution at an arbitrary iteration level $n \neq 0$

$$F_{n0}(x) = \frac{1}{\hat{W}_{(n-1)0}(x)}(\ln U_n(x))', \tag{44}$$

where $U_n(x)$ is some function, the equations Eq.(43) all reduce to Eq.(21). Hence, attempting to generate novel deformations recursively via a product structure, of the kind above, fails. This conclusion was also reached in [6], but at the level of the second order linear differential equation Eq.(21).

3.3. Recursive non-linear deformations

We have so far only considered linear (in the superpotential) deformations. In this section we will briefly consider two non-linear deformation schemes. Let us first consider a polynomial kind of deformation. That is, given a superpotential $\hat{W}_{(i-1)0}(x)$ which we will assume is derived, in some way or another, from some seed superpotential $W(x)$. Consider then the polynomial deformation

$$\hat{W}_{i0}(x) = F_{i0}(x)W^k(x) + \hat{W}_{(i-1)0}(x) ; k \in \{1, 2, 3, \dots\}. \tag{45}$$

The isospectrality condition then implies

$$F'_{i0}(x) + [k(\ln W(x))' + 2\hat{W}_{(i-1)0}(x)]F_{i0}(x) + W^k(x)F^2_{i0}(x) = 0. \tag{46}$$

This is a Riccati type equation of the kind we have met earlier in this work. Apparently, different k-values give rise to very different equations to solve. However, and rather intriguingly, all the possible k-values implies the same deformation. This is seen by making the following substitution

$$F_{i0}(x) = \frac{1}{W^k(x)} \frac{U'_k(x)}{U_k(x)}, \tag{47}$$

where $U_k(x)$ is some function. This expression inserted into Eq.(46) gives

$$U''_k(x) + 2\hat{W}_{i-1}(x)U'_k(x) = 0. \tag{48}$$

Hence, $\hat{W}_{i0}(x)$ is independent of k and we are essentially left with a linear deformation. Clearly, the range of values of k can be expanded to the real numbers.

Another canonical generalization of our work is to consider deformations on the form

$$\hat{W}(x) = H_0(x)\mathcal{F}(W), \tag{49}$$

where $\mathcal{F}$ is any *functional* of the seed superpotential $W(x)$. The isospectrality condition then implies

$$H'_0(x) + (\ln \mathcal{F}(W))'H_0(x) + \mathcal{F}(W)H^2_0(x) = \mathcal{F}(W)^{-1}(W^2(x) + W'(x)). \tag{50}$$

Note that $H_0(x) = 1$ does *not* solve this equation unless $\mathcal{F}(W) = W$, since Eq.(50) with $H_0(x) = 1$ implies $\mathcal{F}'(x) + \mathcal{F}^2(x) = V_+(x)$. Since $V_+(x)$ is uniquely given in Eq.(2) any other choice of functional will fail to satisfy the isospectrality condition. Hence the conclusion. The particular solution $H_0(x) = 1$ is not forced upon us. We can in principle do without it. It is easily verified that Eq.(50) can be cast into the form Eq.(21) by the substitution

$$H_0(x) = \frac{1}{\mathcal{F}(W)}(\ln U(x))'. \tag{51}$$

We can also look for an expanded solution by writing

$$H_0(x) = H_{00}(x) + \frac{1}{Z_0(x)}, \tag{52}$$

where $H_{00}(x)$ is a particular solution of Eq.(50). We then get the equation

$$\frac{d}{dx}Z_0(x) - (\frac{d}{dx}\ln \mathcal{F}(W) + 2H_{00}(x)\mathcal{F}(W))Z_0(x) = \mathcal{F}(W), \tag{53}$$

which is a generalized form of Eq.(18). The reciprocal solution has the general form

$$\frac{1}{Z_0(x)} = \frac{e^{-2\int^x H_{00}(t)\mathcal{F}(W)dt}}{\mathcal{F}(W)(C + \int^x e^{-2\int H_{00}(t)\mathcal{F}(W)dt}du)}. \tag{54}$$

C is an integration constant, which we assume to be real. Utilizing that $H_{00}(x) = \mathcal{F}^{-1}(x)(\ln U(x))'$ the resulting deformation coincides with Eq.(19). We thus therefore conclude that non-linear deformations on the form Eq.(49) does not generate additional deformations to the ones already generated by Eq.(13).

3.4. Deforming the Coulomb potential

As a relatively simple application of the linear deformation scheme let us briefly consider deformations of the Coulomb potential. This potential has, within the framework of SUSYQM, been treated in several previous works [5]. The superpotential and the partner potential for the Coulomb potential are given by [5]

$$W(x) = \frac{q^2}{2(l+1)} - \frac{(l+1)}{x}, \tag{55}$$

$$V_+(x) = \frac{1}{4}\left(\frac{q^2}{l+1}\right)^2 - \frac{q^2}{x} + \frac{(l+1)(l+2)}{x^2}. \tag{56}$$

q and l in these expressions are the electric charge and the angular momentum quantum numbers, respectively. These potentials result in the following general solution for $U_0(x)$ in Eq.(21) [6]

$$U_0(x) = C_1 M_{l+1,l+\frac{3}{2}}\left(\frac{q^2 x}{l+1}\right) + C_2 W_{l+1,l+\frac{3}{2}}\left(\frac{q^2 x}{l+1}\right). \tag{57}$$

The $M(x)$- and $W(x)$-functions are the Whittaker functions. The solution Eq.(22) is given by [6]

$$U_0(x) \sim e^{\frac{q^2 x}{2(l+1)} - (l+1)\ln(2x)}. \tag{58}$$

We will for simplicity assume this solution in the following. We will let $C_{01} \to \infty$ in Eq.(19) such that we deal with the identity deformation $\hat{W}_0(x) = \hat{W}_{00}(x) = W(x)$. We will also ignore the $Q_j(x)$ contributions in the following. Define $A \equiv q^2/(2(l+1))$ and $B \equiv l+1$. It then follows that

$$P_1(x) = C_{11} + C_{12}\int^x t^{2B}e^{-2At}dt, \tag{59}$$

such that

$$\hat{W}_{10}(x) = A - \frac{B}{x} + \frac{C_{12}x^{2B}e^{-2Ax}}{C_{11} + C_{12}\int^x t^{2B}e^{-2At}dt}. \tag{60}$$

Let us consider the s-state with $l = 0$ in order to get a better grasp on the content buried in Eq.(60). We also set $q \equiv 1$. The expression for $\hat{W}_{10}(x)$ then reduces to

$$\hat{W}_{10}(x) = \frac{1}{2} - \frac{1}{x} + \frac{C_{12}x^2 e^{-x}}{C_{11} - C_{12}(x^2 + 2x + 2)e^{-x}} \tag{61}$$

after redefining C_{11} such that the lower integration limit of the integral in Eq.(60) does not appear explicitly in the expression for the potential. We will automatically do such redefinitions in the following when it is appropriate. The corresponding physical potential $\hat{V}_{-1}(x)$ can either be derived from the definition $\hat{V}_{-1}(x) \equiv \hat{W}_{10}^2(x) - \hat{W}_{10}'(x)$ or from Eq.(23) with $\hat{W}_{00}(x) = W(x)$ and C_{01} finite. This is a consequence of a regeneration of Eq.(19) by the recursion scheme which was noted in Figure 1. From the definition it follows that

$$\hat{V}_{-1}(x) = \frac{1}{4} - \frac{1}{x} + \frac{C_{12}x(2x - 4)e^{-x}}{C_{11} - C_{12}(x^2 + 2x + 2)e^{-x}} + \frac{2C_{12}^2 x^4 e^{-2x}}{(C_{11} - C_{12}(x^2 + 2x + 2)e^{-x})^2} . \tag{62}$$

In the special case when we set $C_{11} = 0$ the last term in Eq.(61) becomes independent of the exponentials (and C_{12}) and thus reduces to a pure rational function. The physical potential $\hat{V}_{-1}(x)$ generated by $\hat{W}_{10}(x)$ is then given by

$$\hat{V}_{-1}(x) = \frac{1}{4} - \frac{1}{x} + \frac{4x(x + 2)}{(x^2 + 2x + 2)^2} \equiv V_{-}(x) + \frac{4x(x + 2)}{(x^2 + 2x + 2)^2} . \tag{63}$$

Let us go to the second iteration level starting from the expression for $\hat{W}_{10}(x)$ in Eq.(61) with $C_{11} = 0$, for convenience. It then follows that

$$\hat{W}_{20}(x) = \hat{W}_{1}(x) + \frac{C_{22}x^2 e^x}{C_{21}(x^2 + 2x + 2)^2 + C_{22}(x^2 + 2x + 2)e^x} . \tag{64}$$

Note that when $C_{21} = 0$ we get $\hat{W}_{20}(x) = W(x)$. Hence, the deformation scheme allows in general for the possibility that additional iterations in particular cases may regenerate previous potentials in a nontrivial fashion. The expression for the corresponding physical potential is given by

$$\hat{V}_{-2}(x) = \hat{V}_{-1}(x) + \left[\frac{C_{22}x^2 e^x}{C_{21}(x^2 + 2x + 2)^2 + C_{22}(x^2 + 2x + 2)e^x} \right] \times$$
$$\times \left[-4\left(\frac{1}{x} + \frac{\frac{1}{2}x^2}{x^2 + 2x + 2}\right) + \frac{2(C_{21}(2x + 2) + C_{22}e^x)}{C_{21}(x^2 + 2x + 2) + C_{22}e^x} \right] . \tag{65}$$

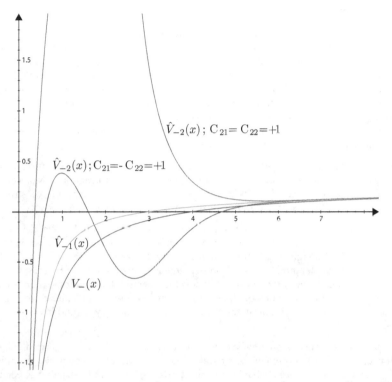

Figure 2. Generic plots depicting $\hat{V}_-(x)$, $\hat{V}_{-1}(x)$ and $\hat{V}_{-2}(x)$. The plots for $\hat{V}_{-2}(x)$ show how drastic the nature of a potential might change as the values of the integration constants change.

The superpotential stemming from the third iteration with $\hat{W}_{20}(x)$ in Eq.(64) as the starting point is given by

$$\hat{W}_{30}(x) = \hat{W}_{20}(x) + \frac{\left(\dfrac{C_{32}x^2e^x}{C_{21}(x^2+2x+2)^2 + C_{22}(x^2+2x+2)e^x}\right)}{\left(C_{31} + \dfrac{C_{32}}{C_{22}}\ln\left|\dfrac{C_{21}(x^2+2x+2) + C_{22}e^x}{x^2+2x+2}\right|\right)}. \tag{66}$$

This superpotential introduces the possibility for a logarithmic singularity away from the origin when $C_{22}/C_{21} < 0$. We note that setting $C_{31} = 0$ does not regenerate a previous potential as was possible at the previous iteration level when we correspondingly put $C_{21} = 0$. From Eq.(66) we can deduce the physical potential $\hat{V}_{-3}(x)$ at the third iteration level. We do not reproduce it here due to its complexity. Due to the complicated integrals appearing we are not able to provide the analytical expression for $\hat{W}_{40}(x)$. We leave detailed studies of the Coulomb potential for the future.

4. Conclusion

In a previous paper we showed that isospectral deformations on the form Eq.(15) are contained in the space of deformations generated by isospectral deformations on the form Eq.(13). In this work we have shown that Eq.(13) can be considered as the initial, or base, deformation of a novel infinite recursive isospectral deformation chain. This thus answers to some extend the question by which we ended our previous paper [6]; how does the most general isospectral deformation of the kind considered there (Eq.(13) in this paper) look like. The results in this work do obviously only give a partial answer. We deduced in particular that a class of recursive deformations exists which is generated by the solutions of the *non-linear* differential equations in Eq.(28).

We briefly discussed various ways to construct alternative recursive deformation structures. We considered a linear product structure, polynomial deformations and completely generalized base deformations. They all either failed to provide a recursive structure or they turned out to be identical to the deformation scheme developed in this work.

We applied the linear recursive scheme to the Coulomb potential. We derived novel superpotentials which all per construction satisfy the isospectrality condition. It is straightforward, although very tedious, to check that the corresponding physical potentials V_+ at the various iteration levels all satisfy the isospectrality condition Eq.(26). This application did also demonstrate how easily novel isospectral deformations can be generated in this approach. It did also demonstrate an increased relative complexity of the generated potentials with the number of iterations, as one also naively would expect from the expression Eq.(33).

The results in this work is obviously only a starting point for further research. One issue which needs clarification is the more general implications which can be drawn from Eq.(28). Another obvious issue is the behaviour of the transmission T and reflection R coefficients when the deformation chain is applied to some known initial scattering process. It is known that T and R are invariant under the simple deformation in Eq.(15) [5]. However, it is unclear whether this property is also a property of the general deformation chain. It is also of great interest to study the relation between our chain construction and the conventional operator approaches. One possible strategy one might follow in order to cast some light on this issue is to study the relation between the deformation chain and the concept of intertwining operators. It is known that many exactly solvable potentials are related by intertwining operator transformations including the Darboux transformations which also appear in our context [6]. Clearly, key to our construction in this work is the non-linearity of the Riccati equation. One could contemplate studying the associated JET-space and its deformations. Furthermore, an analysis of the intertwining of the hierarchy of the JET-spaces associated with the system of equations in Eq.(26) might also cast new light on our subject.

Author details

Bjørn Jensen

* Address all correspondence to: bjorn.jensen@hive.no

Vestfold University College, Norway

References

[1] E. Schrödinger, Proc. R. Irish Acad. **A46**, 9 (1940); **A46**, 183 (1940);   A47, 53 (1941), L.Infeld and T.E. Hull, Rev. Mod. Phys. 23, 21 (1951) (see also the references in this paper).

[2] E. Witten, Nucl. Phys. B188 513 (1981).

[3] G. Darboux, C.R. Acad. Sci. (Paris) 94, 1456 (1882).

[4] D.L. Pursey, Phys. Rev. D33 2267 (1986).

[5] A. Gangopadhyaya , J.V. Mallow and C. Rasinariu , *Supersymmetric quantum mechanics* World Scientific Publishing Co. (2011).

[6] B. Jensen, JHEP 11, 059 (2011).

[7] E. L. Ince,*Ordinary differential equations* Dover Publications, Inc. (1956).

[8] W.E. Boyce, R.C. DiPrima, *Elementary Differential Equations and Boundary Value Problems* New York: Wiley, 4th ed. (1986).

Chapter 8

The Theoretical Ramifications of the Computational Unified Field Theory

Jonathan Bentwich

Additional information is available at the end of the chapter

1. Introduction

Four previous articles (Bentwich, 2012: a-d) have postulated the existence of a novel 'Computational Unified Field Theory' (CUFT) which is a candidate 'Theory of Everything' (Brumfiel, 2006; Ellis, 1986; Greene, 2003) – i.e., has the potential of unifying between quantum (Born, 1954) and relativistic models of physical reality (and moreover also possesses the potential of opening new 'vistas' for scientific examination connected with its discovery of a new hypothetical 'Universal Computational Principle which carries out an extremely rapid computation, c^2/h of a series of Universal Simultaneous Computational Frames, 'USCF's, which give rise to all 'apparent' secondary computational 'physical' properties of 'space', 'time', 'energy' and 'mass'); Indeed, the primary focus of the current manuscript is precisely to explore the potential theoretical ramifications of this novel CUFT – based on the recognition that the (singular) Universal Computational Principle ('ʊ') solely produces all (apparent) secondary computational 'physical' properties of 'space, 'time', 'energy' and 'mass', and hence comprises the sole "reality" (which produces all exhaustive hypothetical inductive and deductive phenomenon through a higher-ordered 'a-causal' computational framework; this may subsequently bear significant theoretical ramifications for all (key) 'material-causal' scientific paradigms as well as point at the discovery of a (new) *'Universal Consciousness Principle Computational Program'*, as well as outline potential resolutions of major Physical 'enigma's;

Hence, the current manuscript traces the potential theoretical ramifications of:

a. An 'a-causal' computational framework of the (CUFT's) singular Universal Consciousness Principle's ('ʊ') responsible for the (higher-ordered) computation of all exhaustive hypothetical (e.g., empirically knowable) inductive or deductive 'x-y' pairs series – which leads to the discovery of a-*causal 'Universal Consciousness Principle Computational Program'*.

b. An exploration of the CUFT's Universal Consciousness Principle's ('ּי') and Duality Principle's (Bentwich, 2003c, 2004, 2006) reformalization of all (apparent inductive or deductive) major SROCS computational paradigms (e.g., including: Darwin's 'Natural Selection Principle' (Darwin, 1859) and associated Genetic Encoding hypothesis, Neuroscience's Psychophysical Problem of human Consciousness and all inductive and deductive Gödel-like SROCS paradigms).

c. Theoretical Ramifications of the Universal Consciousness Principle.

2. A singular 'a-causal' universal consciousness principle computation of all inductive and deductive 'x-y' relationships

We thus begin with an exploration of three potential theoretical ramifications of the CUFT's description of the operation of the (singular) Universal Consciousness Principle ('י') which has been shown to compute an extremely rapid series of Universal Simultaneous Computational Frames (USCF's);

The Universal Computational/Consciousness Principle was (previously) shown to encapsulate a singular higher-ordered 'D2' computation of an 'a-causal' computation of the "simultaneous co-occurrences" of all exhaustive hypothetical inductive or deductive (e.g., empirically knowable) 'x-y' pairs series; Therefore, the acceptance of the CUFT's description of the Universal Consciousness Principle necessarily implies that throughout the various (inductive or deductive) disciplines of Science we need to shift from the current basic (Cartesian) "material-causal" scientific theoretical towards a singular (higher-ordered 'D2') 'Universal Consciousness Principle's a-causal computation':

This means that the current (Cartesian) 'material-causal' scientific framework assumes that any given 'y' element (or value) can be explained as a result of its (direct or indirect) 'causal' interaction/s with another (exhaustive hypothetical inductive or deductive) series of 'x' factor/s – which determines whether that 'y' element (or value) "exists" or "doesn't exist", thereby comprising a 'Self-Referential Ontological Computational System' (SROCS) (Bentwich: 2012a-d):

SROCS: PR{x,y}→ ['y' or 'not y']/di1…din.

But, since it was previously shown that such SROCS computational structure inevitably leads to both 'logical inconsistency' and 'computational indeterminacy' that were shown to be contradicted by robust empirical findings indicating the capacity of the major scientific SROCS paradigms to be capable of determining the "existence" or "non-existence" of the particular 'y' element, see Bentwich 2012b) – then the CUFT's 'Duality Principle' asserted the existence of the singular 'Universal Consciousness Principle' ('י') which is capable of computing the "simultaneous co-occurrences" of any particular (exhaustive hypothetical) 'x-y' pairs series which are embedded within the Universal Computational/Consciousness Principle's rapid series of USCF's.

What this means is that both specifically for each of the (previously identified) key scientific SROCS paradigms as well as more generally for any hypothetical ('empirically knowable') inductive or deductive ('x-y') phenomenon, we must reformulate our scientific understanding in such a way which will allow us to present any such 'x-y' relationship/s as being computed by the singular Universal Consciousness Principle (e.g., as the computation of an exhaustive-hypothetical "co-occurring" 'x-y' pairs' series); In that respect, this (novel) 'Universal Consciousness Principle's' scientific framework shifts Science from its current basic (Cartesian) assumption wherein all natural phenomena can be described as 'material-causal' ('x→y') relationships (e.g., comprising the apparent SROCS computational structure contradicted by the computational Duality Principle) – to an 'a-causal' singular Universal Consciousness Principle which computes the simultaneous "co-occurrences" of any inductive or deductive 'x-y' pairs series comprising the various 'pixels' of the USCF's frames (e.g., produced by this Universal Consciousness Principle).

Finally, it should be noted that a key principle underlying this shift from the current 'material-causal' (Cartesian) scientific framework towards the CUFT's (proven) higher-ordered singular Universal Consciousness Principle's ('ъ') 'a-causal' theoretical framework is the acceptance of the impossibility of the existence of any such 'material-causal' ('x-y') relationship/s – i.e., due to the impossibility of any 'physical' entity, attribute (or property) being transferred across any (two subsequent) 'USCF's frames: Thus, apart from the (previously shown) conceptual computational proof of the 'Duality Principle' wherein due to the inevitable 'logical inconsistency' and 'computational indeterminacy' arising from the SROCS computational structure (which is contradicted by empirical evidence indicating the capacity of these key scientific SROCS paradigms to compute the "existence" or "non-existence" of any particular 'y' element or value) – which points at the existence of the higher-ordered (singular) 'Universal Computational/Consciousness Principle' that computes the "simultaneous co-occurrences" of any (exhaustive-hypothetical) 'x-y' pairs' series; it is suggested that the inclusion of this computational Duality Principle as one of the (seven) theoretical postulates of the CUFT (e.g., specifically alongside the CUFT's 'Computational Invariance' and 'Universal Consciousness' postulates) unequivocally asserts that there cannot (in principle) exist any 'material-causal' effect/s (or relationship/s) being transferred across any (two subsequent) USCF's frames! This is because the CUFT's very definition of all four 'physical' properties of 'space', 'time', 'energy' and 'mass' – as secondary computational by-products of the (singular) Universal Computational Consciousness' computation of (an extremely rapid series of) 'Universal Simultaneous Computational Frames' (USCF's); and moreover the CUFT's 'Computational Invariance' postulate indication that due to the 'computational variance' of these four (secondary computational) 'physical' properties (e.g., as existing only "during" the appearance of the USCF frames but 'non-existence' "in-between" any two such subsequent frames, see Bentwich, 2012:c-d) as opposed to the 'computational invariance' of the 'Universal Consciousness Principle' ('ъ'), we need to regard only this singular (computationally invariant) 'Universal Consciousness Principle' as "real" whereas all four (secondary computationally variant) 'physical' properties must be regarded as merely 'phenomenal' (i.e., as being comprised in reality only from the singular Universal Consciousness Principle); Therefore, the CUFT's 'Universal Consciousness Principle' advocated that none of these four (secondary computationally variant) 'physical' properties (e.g., of 'space', 'time', 'energy' or 'mass') "really" exists – but rather that there is only this one singular Universal Consciousness Principle which exists (solely) "in-between" any (two subsequent) USCF's frames and also solely produces each of these USCF's derived four 'phenomenal physical' properties; Hence, it was evinced (by the CUFT's Universal Consciousness Principle) that there cannot be any 'transference' of any hypothetical 'material' or 'physical' entity, effect, or property across any (two subsequent) USCF's frames! We therefore reach the inevitable theoretical conclusion that the current scientific (Cartesian) "material-causality' basic assumption underlying all key scientific SROCS paradigms as well as all (empirically knowable) 'Gödel-like' (inductive or deductive) SROCS 'x-y' relationships, wherein there exists a 'material-causal' effect/s (or relationship/s) between any given 'x' element and any (exhaustive hypothetical) 'y' series which determines the "existence" or "non-existence" of that (particular) 'y' element (or value) – is untenable! Instead, we must accept the CUFT's assertion that there can only exists one singular 'Universal Consciousness Principle' ('˙') which both (solely) produces- all (apparent) secondary computational 'physical' properties (of 'space', 'time', 'energy' and 'mass'), as well as computes the "simultaneous co-occurrences" of any (particular) exhaustive-hypothetical inductive or deductive 'x-y' pairs series (e.g., comprising the exhaustive USCF's frames)

3. The "universal consciousness principle's computational program"

Therefore, it follows that based on the recognition of the singularity of the Universal Consciousness Principle's 'a-casual' computation of the "simultaneous co-occurrences" of all (inductive or deductive) 'x-y' pairs' series (as comprising the exhaustive USCF's frames) – we need to be able to reformulate all of the previously mentioned key scientific SROCS paradigms (Bentwich, 2012b-d), including: Darwin's 'Natural Selection Principle' and associated 'Genetic Encoding' hypothesis, Neuroscience's Psychophysical Problem of human Consciousness, as well as all (exhaustive hypothetical) 'Gödel-like' (apparent) inductive or deductive SROCS computational paradigms based on this singular (higher-ordered) Universal Consciousness Principle's ('˙') 'a-causal' USCF's computation;

Hence, what follows is a description of the principle theoretical ramifications of reformulating each of these key scientific (apparent) SROCS computational paradigms, as well as a more generalized description of a tentative 'Universal Consciousness Principle Program' (e.g., which may offer a successful alternative for 'Hilbert's Mathematical Program' to base all of our human scientific knowledge upon the foundations of the operation of the singular Universal Consciousness Principle). First, it may be worthwhile to rearticulate the reformalization of each of these key scientific (apparent) SROCS paradigms in terms of the operation of the singular Universal Consciousness Principle (as previously outlined: Bentwich, 2012b):

$S.:$**D2**: $[\{E\{1...n\}, o\}st1; \{E\{1...n\}, o\}st2... \{E\{1...n\}, o\}stn]$.

$G.F - P.S.:$**D2**: $[\{G\{1...n\}, \text{'ph}i\ (o)'\}st1; \{G\{1...n\}, \text{'ph}j\ (o)'\}sti;...\{G\{1...n\}, \text{'ph}n(o)'\}stn]$.

$G.E. - P.S.:$**D2**: $[\{Ge\{1...n\}, pi\text{-}synth\ (o\text{-}phi)\}st1; Ge\{1...n\}, pj\text{-}synth\ (o\text{-}phi)\}sti... ; Ge\{1...n\}, pn\text{-}synth\ (o\text{-}phi)\}stn]$

$Psychophysical:$**D2**: $[\{N(1...n)_{st\text{-}i}, Cs\text{-}pp_{st\text{-}i}\}; ... \{N(1...n)_{st\text{-}i+n}, Cs\text{-}pp_{st\text{-}i+n}\}]$

Functional: **D2**: [{Cs(pp)f$_i$, Na(spp)fi}$_{st-i}$; ... {Cs(pp)f$_{(i+n)}$, Na(spp)f$_{(i+n)}$}$_{st(i+n)}$]

Phen.: **D2**: [{Cs(pp- fi)-Ph$_i$, Na(spp-fi)-Ph$_i$}$_{st-i}$; ...{Cs(pp- fi)-Ph$_{(i+n)}$, Na(spp-fi)-Ph}$_{st-(i+n)}$]

Self: **D2**: [{Cs(pp- fi)Ph-Si, Na(pp- fi)Ph-S$_i$}$_{st-i}$; ...{Cs(pp- fi)Ph-S(i+n), Na(pp- fi)Ph-S(i+n)}$_{st-(i+n)}$]

GIT:**D2**: ([{S{1...n}, t}i ... {S{1...n}, t}z], or [{x{1...n}i, yi} ... {x{1...n}z, yz}])

Indeed, what may be seen from this singular description of all of these key scientific SROCS paradigms, is that it recognize the fact that all of these major (apparent) SROCS paradigms are computed simultaneously as different "co-occurring" 'x-y' pairs embedded within the same (single or multiple) USCF frame that is produced by the singular Universal Consciousness Principle ('י'); What this means is that the recognition of the singularity of this Universal Consciousness Principle as the sole "reality" which computes the "simultaneous co-occurrences" of all of these (particular) exhaustive hypothetical 'x-y' pairs series, and which also exists (solely) "in-between" any two such USCF's – forces us to transcend the 'narrow constraints' of the (current) Cartesian 'material-causal' theoretical framework (e.g., which assumes that any given 'y' entity (or phenomenon) is "caused" by its (direct or indirect) physical interaction/s with (an exhaustive hypothetical 'x' series); Instead, this singular Universal Consciousness Principle 'a-causal' computation asserts that it is the same singular Universal Consciousness Principle which computes- produces- retains- and evolves- all of these particular scientific (apparent) SROCS 'x-y' pairs series across a series of USCF's...

In other words, instead of the existence of any "real" material-causal relationship between any of these (particular SROCS) 'x→y' entities (e.g., Darwin's Natural Selection Principle's assumed 'material-causal' relationship between an organism's Environmental Factors, 'x', and own traits or behavior 'y'; or between any exhaustive hypothetical Genetic Factors and any given phenotypic behavior; or between Neuroscience's Psychophysical Problem of Human Consciousness' psychophysical stimulation, 'x', and Neural Activation, 'y'; or in fact between any hypothetical inductive or deductive Gödel-like SROCS 'x-y' factors); the CUFT's Universal Consciousness Principle offers an alternative singular (higher-ordered) computational mechanism which computes the "simultaneous co-occurrences" of any of these (exhaustive hypothetical) 'x-y' pairs' series – which are all produced- and embedded- within the Universal Consciousness Principle's computed USCF's frames... Indeed, the shift from the current 'material-causal' (Cartesian) scientific framework towards the Universal Consciousness Principle's singular computation of the "simultaneous co-occurrences" of all exhaustive hypothetical (inductive or deductive) 'x-y' pairs' series may lead the way for reformulating all of these key scientific SROCS paradigms (as well as any other hypothetical inductive or deductive 'x-y' series) within a basic "Universal Consciousness Principle Computational Program";

Essentially, such a *'Universal Consciousness Principle's Computational Program'* is based upon the foundations of the CUFT's (abovementioned) three postulates of the 'Duality Principle', the 'Computational Invariance' principle and the 'Universal Consciousness Principle' – all pointing at the fact that all empirically computable (inductive or deductive) 'x-y' relationships must necessarily be based upon the singular (conceptually higher-ordered) Universal Consciousness Principle which is solely responsible for the computation of the "simultaneous co-occurrences" of all such (exhaustive hypothetical) inductive or deductive 'x-y' pairs series comprising the totality of the USCF's (single or multiple) frames.... Moreover, this singular Universal Consciousness Principle ('י') was also shown to exist independently of any (secondary computational) 'physical properties' (e.g., of 'space', 'time', 'energy' and 'mass') and therefore constitute the only "reality" that exists invariantly (i.e., both as giving rise to the four 'phenomenal' physical properties and as existing solely "in-between" any two such subsequent USCF's frames).

In order to appreciate the full (potential) theoretical significance of such a 'Universal Consciousness Principle Computational Program' it may be worthwhile to reexamine Hillbert's famous 'Mathematical Program' to base Mathematics upon the foundations of Logic (e.g., and by extension also all of Science upon the foundations of Mathematics and Logic), and more specifically, to revisit 'Gödel's Incompleteness Theorem' (GIT) which delivered a critical blow to Hilbert's 'Mathematical Program'; It is a well-known that Hilbert's Mathematical Program sought to base Mathematics (e.g., and by extension also the rest of inductive and deductive Science) upon a logical foundation (e.g., of certain axiomatic definitions); It is also well known that Gödel's Incompleteness Theorem (GIT) has failed Hillbert's Mathematical Program due to its proof that there exists certain 'self-referential' logical-mathematical statements that cannot be determined as "true" or "false" (e.g., or logically 'consistent' or 'inconsistent') from within any hypothetical axiomatic logical-mathematical system. Previously (Bentwich, 2012c,d) it was suggested that perhaps scientific Gödel -like SROCS computational systems may in fact be constrained by the Duality Principle's (generalized) format, thus:

i.   SROCS: PR{x,y}→['y' or 'not y']/di1...din

ii.   SROCS CR{S,t}→ ['t' or 'not t']/di1...din

wherein it was shown that both inductive ('i') and deductive (ii) SROCS scientific computational systems are necessarily constrained by the Duality Principle (e.g., as part of the broader CUFT). In other words, the Duality Principle's (generalized format) was shown to constrain all (exhaustive hypothetical) Gödel -like (inductive or deductive) scientific SROCS paradigms, thereby pointing at the existence of a singular (higher-ordered) Universal Consciousness Principle ('י') which is solely capable of computing the "simultaneous co-occurrences" of any (exhaustive hypothetical) 'x-y' pairs series. It is important to note, however, that the conceptual computational constraint imposed upon all (Gödel -like) inductive or deductive scientific SROCS paradigms was shown to apply for all of those inductive or deductive (apparent) scientific SROCS paradigms – for which there is an empirically known (or 'knowable') 'x-y' pairs series results!

This latter assertion of the Duality Principle's (generalized proof) may be significant as it both narrows- and emphasized- the scope of the 'scientifically knowable domain'; In other words, instead of the current 'materialistic-reductionistic' scientific framework which is anchored in a basic (inductive or deductive) SROCS computational format (see above) which inevitably leads to both 'logical inconsistency' and 'computational indeterminacy' that are contradicted by robust empirical findings (e.g., pertaining to the key scientific SROCS paradigms); The Duality Principle (e.g., as one of the postulates within the broader CUFT) proves that the only means for computing the "simultaneous co-occurrences" of any (exhaustive hypothetical) 'x-y' pairs series is carried out by the singular (higher-ordered) Universal Consciousness Principle ('ᕐ')... Moreover, the (generalized format of the) Duality Principle goes farther to state that for all other (exhaustive hypothetical) inductive or deductive computational SROCS paradigms – *for which there exists a proven empirical capacity to determine the values of any particular 'x-y' pairs (e.g., empirically "known" or "knowable" 'x-y' pairs results)*- any of these (hypothetical) scientific SROCS computations must be carried out by the CUFT's identified singular Universal Consciousness Principle ('ᕐ')!

The (potential) significance of this generalized assertion made by the Computational Unified Field Theory's (CUFT): 'Duality Principle', 'Computational Invariance' principle and Universal Consciousness Principle ('ᕐ') is twofold:

a.   First, it narrows down the scope of (inductive or deductive) determinable scientific phenomena – to only those (inductive or deductive) 'x-y' relationships for which there is an empirical capacity to determine their "simultaneously co-occurring" values; essentially the *'Universal Consciousness Principle's Computational Program'* anchors itself in the Duality Principle's focus on only those inductive or deductive 'x-y' relationship/s or phenomenon for which there is an empirically 'known' or 'knowable' capacity to determine these 'x-y' pairs values. It is perhaps important to note (in this context) that all of the 'other' inductive or deductive 'x-y' relationship/s which cannot be (empirically) known – "naturally" lie outside the scope of our human (scientific) knowledge (and therefore should not be included, anyway within the scope of Science)... Nevertheless, the strict limitation imposed by the 'Universal Consciousness Principle Computational Program' – may indeed be significant, as it clearly defines the boundaries of "admissible scientific knowledge" to only that scientific knowledge which is based on empirically known or knowable results pertaining to the "simultaneous co-occurrences" of any 'x-y' relationship or phenomenon; (Needless to say that the strict insistence of the Universal Consciousness Computational Program upon dealing only with

b.   Second, based on this strict definition of Science as dealing solely with 'empirical knowable' (simultaneously co-occurring) 'x-y' relationship/s or phenomenon – the 'Universal Consciousness Computational Program' may in fact offer a broader alternative to GIT (failing of Hilbert's 'Mathematical Program'); This is because once we accept the Universal Consciousness Principle's Computational Program's (above) strict 'empirical constrains', we are led to the Duality Principle's (generalized) conceptual computational proof that any (exhaustive hypothetical) inductive or deductive scientific SROCS' 'x-y' relationship must be determined by the singular Universal Consciousness Principle ('ᕐ') computation of the "simultaneous co-occurrences" of any (exhaustive hypothetical) 'x-y' pairs series. This means that instead of GIT assertion that it is not possible (in principle) to construct a consistent Logical-Mathematical System which will be capable of computing any mathematical (or scientific) claim or theorem, the Universal Consciousness Computational Program asserts that based on a strict definition of Science as dealing solely with empirically knowable 'x-y' relationship/s or phenomenon, we obtain a singular (higher-ordered) Universal Consciousness Principle which is solely responsible for computing the "simultaneous co-occurrences" of any (exhaustive hypothetical) inductive or deductive 'x-y' pairs series (e.g., which were shown by the CUFT to comprise the totality of any single or multiple USCF's frames that are solely produced by this Universal Consciousness Principle). In that sense, it may be said that the Universal Consciousness Principle Computational Program points at the existence of the singular (higher-ordered) Universal Consciousness Principle as constraining- and producing- all inductive or deductive scientific relationship/s or phenomena (e.g., which was also shown earlier and previously to constitute the only "reality" which both produces all USCF's derived secondary computational 'physical properties and also solely exists "in-between" any two such USCF's).

4. Theoretical ramifications of the universal consciousness principle

The discovery of the singular Universal Consciousness Principle (alongside its 'Universal Consciousness Computational Program') may bear a few significant theoretical ramifications:

1. **The Sole "Reality" of the Universal Consciousness Principle:** As shown above, all scientific (inductive and deductive) disciplines need to be reformulated based on the recognition that there exists only a singular (higher-ordered) Universal Consciousness Principle ('ᴗ') which solely produces- sustains- evolves (and constrains) all (apparent) SROCS (inductive or deductive) 'x-y' relationships; Moreover, this Universal Consciousness Principle is recognized as the sole "reality" that both produces- sustains- and evolves- any of the apparent (four) 'physical' properties of 'space', 'time', 'energy' and 'mass', as well as exists independently of any such 'physical' properties – and is therefore recognized as the only singular "reality", whereas these apparent 'physical' properties are seen as merely 'phenomenal' (secondary computational) manifestations of this singular (higher-ordered) Universal Consciousness Principle "reality".

2. **The Transcendence of 'Material-Causality' by the Universal Consciousness Principle 'A-Causal' Computation:** As shown (above), the acceptance of the Universal Consciousness Principle ('ᴗ') as the sole "reality" which both produces- (sustains- and evolves-) all USCF's (secondary computational) 'physical' properties, as well as exists independently "in-between" any (two subsequent) USCF's; (Alongside the Duality Principle's negation of any apparent SROCS' 'causal' relationships and the 'Computational Invariance' principle indication that only the 'computationally invariant' 'Universal Consciousness Principle' "really" exists whereas the secondary 'computationally variant' physical properties are only 'phenomenal'.) – point at the negation of any "real" material-causal ('x-y') relationships, but instead indicate that there can only exist a singular (higher-ordered) Universal Consciousness Principle 'a-causal' computation of the "simultaneous co-occurrences" of any exhaustive hypothetical inductive or deductive 'x-y' pairs' series... (As shown earlier, the strict negation of the existence of any "real" 'material-causal' 'x→y' relationships was evinced by the simple fact that according to the CUFT's model there cannot exist any "real" computationally variant 'physical' or 'material' property that can

"pass" across any two subsequent USCF's, but only the computationally invariant "real" Universal Consciousness Principle which exists singularly – as solely producing all apparent secondary computational 'physical' properties as well as existing independently "in-between" any two such subsequent USCF's frames.) Indeed, the need to replace all apparent 'material-causal' 'x-y' SROCS relationships by a singular (higher-ordered) Universal Consciousness Principle computation of the 'simultaneous co-occurrences' of all possible inductive or deductive 'x-y' pairs series was shown to apply to all of the key (apparent) scientific SROCS paradigms (including: Darwin's Natural Selection Principle and associated Genetic Encoding hypothesis, Neuroscience's Psychophysical Problem of human Consciousness as well as to all Gödel-like hypothetical inductive or deductive SROCS paradigms; what this implies is that for all of these apparent SROCS scientific paradigms the sole "reality" of the Universal Consciousness Principle forces us to tran-scend each of the (particular) 'material-causal' x-y relationships in favor of the Universal Consciousness Principle's singular computation of all (exhaustive hypothetical) 'x-y' pairs series; Thus, for example, instead of Darwin's current 'Natural Selection Principle' SROCS material-causality thesis, which assumes that it is the direct (or indirect) physical inter-action between the organism and its Environmental Factors that causes that organism to 'survive' or be 'extinct', the adoption of the Universal Consciousness Principle (and Duality Principle) postulates brigs about a recognition that there is only a singular (Universal Consciousness based) conceptually higher-ordered 'a-causal' computation of the "simul-taneous co-occurrences" of an exhaustive hypothetical pairs series of 'organism' and 'Environmental Factors' (e.g., which are computed as part of the Universal Consciousness Principle's production of the series of USCF's frames).

3. **Possible Resolution of Physical (and Mathematical) Conundrums:** It is suggested that certain key Physical (and Mathematical) Conundrums including: Physics "dark energy", "dark matter" and "arrow of time" enigmas may be potentially resolved through the application of this singular 'Universal Consciousness Principle'; this is because according to the CUFT, all (four) 'physical' properties of 'space', 'time', 'energy' and 'mass' are (in reality) solely produced by the Universal Consciousness Principle (e.g., as secondary computational 'phenomenal' properties); Hence, the key enigma of "dark energy" and "dark matter" (e.g., the fact that based on the calculation of the totality of 'mass' and 'energy' in the observable cosmos the expansion of the universe should not be as rapid as is observed – which is currently interpreted as indicating that approximately 70-90% of the "energy" and "mass" in the universe in "dark", that is not yet observable) – may be explainable based on the CUFT's delineation of the Universal Consciousness Principle's (extremely rapid) computation of the series of USCF's. This is due to the fact that according to the Universal Consciousness Principle's (previously discovered: Bentwich, 2012a) 'Universal Computational Formula' the production of any "mass" or "energy" ("space" or "time") 'physical' properties – are entirely (and solely) produced through the Universal Consciousness Principle's computation of the degree of 'Consistency' (e.g., 'consistent' or 'inconsistent') across two other Computational Dimensions, i.e., 'Framework' ('frame' vs. 'object') and 'Locus' ('global' vs. 'local'): Thus, for instance it was shown that any "mass" measurement of any object in the universe is computed by the Universal Consciousness

Principle ('·') as the degree of 'consistent-object' measurement (of that particular) object across a series of USCF frames.

Hence, by extension, the totality of the "mass" measured across the entire physical universe should be a measure of the degree of consistent-object/s values across a series of USCF's! Note, however, that based on the abovementioned recognition that in "reality" – only the Universal Consciousness Principle ('·') "exists" (e.g., both as producing any of the USCF's derived four secondary computational 'physical' properties as well as existing independently "in-between" any two such USCF's frames), and therefore that only this Universal Consciousness Principle "really" produces all of the (apparent) "mass" and "energy" in the 'physical' universe (e.g., rather than the "energy" and "mass" in the 'physical' universe being "caused" by the "material" objects in the cosmos)... Hence, also all of the "energy" in the physical universe is solely produced by this (singular) Universal Consciousness Principle, e.g., as a measure of the degree of 'inconsistent-frame' (changes) of all of the objects (I the universe) across a series of USCF's frames. Therefore, according to the CUFT, the explanation of all of the "mass" and "energy" values observed in the 'physical' universe – should be solely attributed to the operation of the Universal Computational Principle, i.e., through its (extremely rapid) computation of the rapid series of USCF's (respective secondary computational measures of the abovementioned degree of 'consistent-object': "mass", or 'inconsistent-frame': "energy"). We therefore obtain that the (accelerated) rate of expansion of the physical universe – should be explained (according to the CUFT) based on the Universal Consciousness (extremely rapid) computation of the USCF's (e.g., which gives rise to the apparent secondary computational 'physical' measures of 'consistent-object': "mass" or 'inconsistent-frame': "energy"), rather than arise from any 'material-causal' effects of any (strictly hypothetical) "dark mass" or "dark energy"... (Once again, it may be worth pointing at the abovementioned conceptual computational proof that there cannot be any transference of any "physical" property (entity or effect etc.) across any (two subsequent) USCF's frames, but only the retention- or evolution- of all of the spatial pixels' "physical" properties by the singular Universal Consciousness Principle across the series of USCF's – which therefore also precludes the possibility of any "real" "material" effects exerted by any "dark" mass or energy on the expansion of the 'physical' universe across a series of USCF frames.)

Similarly, the "arrow of time" conundrum in modern Physics, e.g., which essentially points at the fact that according to the laws of Physics, there should not be any difference between the physical pathways of say the "breaking of a glass cup into a (thousand) small glass' pieces" and the "re-integration of these thousand glass' pieces into a unitary glass cup"! In other words, according to the strict laws of Physics, there should not be any preference for us seeing "glasses" break into a thousand pieces – over our seeing of the thousand pieces become "reintegrated" into whole glass cups (again), which is obviously contradicted by our (everyday) phenomenal experiences (as well as by our empirical scientific observations)... Hence, according to the current state of (quantum and relativistic) models of Physical reality – there is no reasonable explanation for this "arrow of time" apparent empirical "preference" for the "glass breaking into pieces" scenario over the "reintegration of the glass pieces" scenario...

However, it is suggested that according to one of the CUFT critical empirical predictions (previously outlined: Bentwich, 2012b) this "arrow of time" Physical conundrum may be resolved: This is because one (of three) critical empirical predictions of the CUFT assert the possibility of reversing any spatial-temporal sequence associated with any given 'electromagnetic spatial pixel' through the appropriate manipulation of that object's (or event's) electromagnetic spatial pixel values (across a series of USCF's): It was thus indicated that if we were to accurately record the spatial electromagnetic pixels' values of any particular object (e.g., such as an amoeba or any other living organism for instance) across a series of USCF's frames (e.g., or even through a certain sampling from a series of USCF's), and to the extent that we could appropriately manipulate these various electromagnetic spatial pixels' values in such a manner which allows us to reproduce that objects' electromagnetic spatial pixels' values (across the measured series of USCF's) – in the reversed spatial-temporal sequence, then it may be possible to reverse the "flow of time" (e.g., spatial-temporal electromagnetic pixels' sequence). In this way it should be possible (according to one of the critical predictions of the CUFT) to actually "reverse" the "arrow of time" (e.g., at least for particular object/s or event/s: such as for instance, bring about a situation in which a "broken glass cup may in fact be reintegrated"...)

Author details

Jonathan Bentwich*

Brain Perfection LTD, Israel

References

[1] Bentwich, J. (2003a). From Cartesian Logical-Empiricism to the'Cognitive Reality': A Paradigmatic Shift, *Proceedings of Inscriptions in the Sand, Sixth International Literature and Humanities Conference*, Cyprus

[2] Bentwich, J. (2003b). The Duality Principle's resolution of the Evolutionary Natural Selection Principle; The Cognitive 'Duality Principle': A resolution of the 'Liar Paradox' and 'Gödel's Incompleteness Theorem' conundrums; From Cartesian Logical-

[3] Empiricism to the 'Cognitive Reality: A paradigmatic shift *Proceedings of 12th International Congress of Logic, Methodology and Philosophy of Science*, August Oviedo, Spain

[4] Bentwich, J. (2003c). The cognitive'Duality Principle': a resolution of the'Liar Paradox' and'Gödel's Incompleteness Theorem' conundrums, *Proceedings of Logic Colloquium*, Helsinki, Finland, August 2003

[5] Bentwich, J. (2004). The Cognitive Duality Principle: A resolution of major scientific conundrums, *Proceedings of The international Interdisciplinary Conference*, Calcutta, January

[6] Bentwich, J. (2006). The 'Duality Principle': Irreducibility of sub-threshold psycho-physical computation to neuronal brain activation. *Synthese*, , 153(3), 451-455.

[7] Bentwich, J. (2012a). Quantum Mechanics / Book 1 (979-9-53307-377-3Chapter title: The'Computational Unified Field Theory' (CUFT): Harmonizing Quantum Mechanics and Relativity Theory.

[8] Bentwich, J. (2012b). Quantum Mechanics / Book 1 (979-9-53307-377-3Chapter 23, Theoretical Validation of the Computational Unified Field Theory., 551-598.

[9] Bentwich, J. (2012c). Quantum Mechanics (In Press) Chapter title: The'Computational Unified Field Theory' (CUFT): A Candidate Theory of Everything.

[10] Born, M. (1954). The statistical interpretation of quantum mechanics, *Nobel Lecture, December 11, 1954*

[11] Brumfiel, G. (2006). Our Universe: Outrageous fortune. *Nature*, , 439, 10-12.

[12] Darwin, C. (1859). *On the Origin of Species by Means of Natural Selection, or the Preservation of Favoured Races in the Struggle for Life* John Murray, London; modern reprint Charles Darwin, Julian Huxley (2003). *On The Origin of Species*. Signet Classics. 0-45152-906-5

[13] Ellis, J. (1986). The Superstring: Theory of Everything, or of Nothing? *Nature*, , 23 (6089), 595-598.

[14] Greene, B. (2003). *The Elegant Universe*, Vintage Books, New York

Chapter 9

Shannon Informational Entropies and Chemical Reactivity

Nelson Flores-Gallegos

Additional information is available at the end of the chapter

1. Introduction

During the last decade, researchers around the world have shown that Information Theory [7] is probably one of the most important models in modern science. This model has given rise to applications and reinterpretations of concepts in Physics, Chemistry, Biology, Mathematics, Telecommunications and many other areas that are not, in principle, related to Information Theory. In the 90's, E. R. Frieden showed that important results such as the Schrödinger equation, the Maxwell-Boltzmann and Boltzmann distributions, the Dirac equation can be derived from principles of Information Theory [1–5].

Indded, information is a general concept that is perfectly applicable to any case. It is possible to ask what is the importance of the concept of information measure in quantum mechanics? What do they have in common the codes used to send messages from a communications' satellite have in common with the bases in a DNA molecule? How does the second law of Thermodynamics relate to Communication, to the extent that it is possible to speak of the entropy of a musical theme? How do the intricate problems of probability relate to the way we express ourselves orally or in writing? The answer to these questions can be found in *information*, and the fact that this concept can link very different ideas reveals its great generality and power.

In Chemistry, Information Theory has been applied to the characterization of chemical systems and chemical processes [10–13]. It has been shown that it is possible to use informational entropies to characterize processes such as bond breaking and bond formation. Shannon's entropy can be regarded as a general measure of information that can be used to obtain the Fukui function, which is a parameter of chemical reactivity in atomic and simple molecular systems [6]. In practice, the formal development of applications of Shannon's entropies in Density Functional Theory (DFT) is a fertile area of research. In this context the maximum entropy method has been applied to DFT [33]. Following this line of research, we have initiated a new investigation to derive formal relationships between Information Theory concepts and Theoretical Chemistry.

In this work, we consider the idea that Shannon's entropy can be directly related to some fundamental DFT concepts. To show this, we present some simple mathematical derivations to prove that the first derivative of the Shannon entropy is directly related to DFT reactivity parameters such as the Fukui function, the hardness, softness, and chemical potential and that it might even be possible to obtain a formal relationship between Shannon's entropy and the electron energy. Finally, chemical applications are presented in which the relationships obtained in this work are used in two case studies involving a simple chemical reaction and a conformational analysis.

2. The first derivatives of Shannon's entropy and their relationship to chemical reactivity

For purposes of this chapter we take as a starting point the definition of the unnormalized Shannon's entropy in terms of the electron density in position space, which is defined as:

$$s(\mathbf{r}) = -\int \rho(\mathbf{r}) \ln \rho(\mathbf{r}) d\mathbf{r}, \tag{1}$$

If normalized electron densities are used, $\frac{\rho(\mathbf{r})}{N}$, Shannon's entropy becomes positive: $S(\mathbf{r}) \geq 0$ and the normalized Shannon entropy is

$$S(\mathbf{r}) = -\int \frac{\rho(\mathbf{r})}{N} \ln \frac{\rho(\mathbf{r})}{N} d\mathbf{r}, \tag{2}$$

where N is the number of electrons. This expression can be rewritten as:

$$
\begin{aligned}
S(\mathbf{r}) &= -\frac{1}{N} \int \rho(\mathbf{r}) \ln \left(\frac{\rho(\mathbf{r})}{N} \right) d\mathbf{r} \\
&= -\frac{1}{N} \int \rho(\mathbf{r}) \left[\ln \rho(\mathbf{r}) - \ln N \right] d\mathbf{r} \\
&= -\frac{1}{N} \int \rho(\mathbf{r}) \ln \rho(\mathbf{r}) d\mathbf{r} + \frac{1}{N} \ln N \underbrace{\int \rho(\mathbf{r}) d\mathbf{r}}_{=N} \\
&= -\frac{1}{N} \int \rho(\mathbf{r}) \ln \rho(\mathbf{r}) d\mathbf{r} + \ln N \\
&= -\frac{s(\mathbf{r})}{N} + \ln N.
\end{aligned}
\tag{3}
$$

Differentiating the entropy $S(\mathbf{r})$ with respect to N at constant external potential $v(\mathbf{r})$:

$$\left(\frac{\partial S(\mathbf{r})}{\partial N}\right)_{v(\mathbf{r})} = -\frac{s(\mathbf{r})}{N^2} - \frac{1}{N}\int\left(\frac{\partial \rho(\mathbf{r})}{\partial N}\right)_{v(\mathbf{r})}\ln\rho(\mathbf{r})d\mathbf{r} + \frac{1}{N}, \tag{4}$$

The term $\left(\frac{\partial \rho(\mathbf{r})}{\partial N}\right)_{v(\mathbf{r})}$, is the Fukui function, which is a chemical reactivity parameter in DFT [33].

Also, we can set up the follow relationship

$$\left(\frac{\partial \rho(\mathbf{r})}{\partial \mu}\right)_{v(\mathbf{r})} = \left(\frac{\partial \rho(\mathbf{r})}{\partial N}\right)_{v(\mathbf{r})}\left(\frac{\partial N}{\partial \mu}\right)_{v(\mathbf{r})}, \tag{5}$$

and obtain the variation of Shannon's entropy with respect to the chemical potential:

$$\left(\frac{\partial S(\mathbf{r})}{\partial N}\right)_{v(\mathbf{r})}\left(\frac{\partial N}{\partial \mu}\right)_{v(\mathbf{r})} = -\frac{s(\mathbf{r})}{N^2}\left(\frac{\partial N}{\partial \mu}\right)_{v(\mathbf{r})} - \frac{1}{N}\int\left(\frac{\partial \rho(\mathbf{r})}{\partial N}\right)_{v(\mathbf{r})}\left(\frac{\partial N}{\partial \mu}\right)_{v(\mathbf{r})}\ln\rho(\mathbf{r})d\mathbf{r}$$
$$+\frac{1}{N}\left(\frac{\partial N}{\partial \mu}\right)_{v(\mathbf{r})} \tag{6}$$

simplify,

$$\left(\frac{\partial S(\mathbf{r})}{\partial \mu}\right)_{v(\mathbf{r})} = \left[\frac{1}{N} - \frac{s(\mathbf{r})}{N^2}\right]\left(\frac{\partial N}{\partial \mu}\right)_{v(\mathbf{r})} - \frac{1}{N}\int\left(\frac{\partial \rho(\mathbf{r})}{\partial \mu}\right)_{v(\mathbf{r})}\ln\rho(\mathbf{r})d\mathbf{r}, \tag{7}$$

In this expression, we recognize the term $\left(\frac{\partial \rho(\mathbf{r})}{\partial \mu}\right)_{v(\mathbf{r})}$ which is the local softness.

Also, is it interesting to obtain the variation of Shannon's entropy with respect to the electron density,

$$\left(\frac{\partial S(\mathbf{r})}{\partial \rho(\mathbf{r})}\right)_{v(\mathbf{r})} = -\frac{1}{N}\int 1 + \ln\rho(\mathbf{r})d\mathbf{r}. \tag{8}$$

A similar procedure can be used to obtain the above relations in momentum space as well as for the total Shannon entropy [14], which is defined as

$$S_T = S(\mathbf{r}) + S(\mathbf{p}), \tag{9}$$

where $S(\mathbf{r})$ and $S(\mathbf{p})$ are Shannon's entropies in position and momentum spaces, respectively. The variation of the total Shannon entropy with respect to electron number would be:

$$\frac{dS_T}{dN} = \left(\frac{\partial S(\mathbf{r})}{\partial N}\right)_{v(\mathbf{r})} + \left(\frac{\partial S(\mathbf{p})}{\partial N}\right)_{v(\mathbf{p})}, \tag{10}$$

this permit us, open a door, to the study of this kind of derivatives and chemical descriptors, in momentum space. Results about, the chemical reactivity parameters in momentum space, will be present in other work.

Summarizing, the relations obtained are:

i)

$$\left(\frac{\partial S(\mathbf{r})}{\partial N}\right)_{v(\mathbf{r})} = -\frac{s(\mathbf{r})}{N^2} - \frac{1}{N}\int\left(\frac{\partial \rho(\mathbf{r})}{\partial N}\right)_{v(\mathbf{r})} \ln \rho(\mathbf{r})d\mathbf{r} + \frac{1}{N}. \tag{11}$$

ii)

$$\left(\frac{\partial S(\mathbf{r})}{\partial \mu}\right)_{v(\mathbf{r})} = \left[\frac{1}{N} - \frac{s(\mathbf{r})}{N^2}\right]\left(\frac{\partial N}{\partial \mu}\right)_{v(\mathbf{r})} - \frac{1}{N}\int\left(\frac{\partial \rho(\mathbf{r})}{\partial \mu}\right)_{v(\mathbf{r})} \ln \rho(\mathbf{r})d\mathbf{r}. \tag{12}$$

iii)

$$\left(\frac{\partial S(\mathbf{r})}{\partial \rho(\mathbf{r})}\right)_{v(\mathbf{r})} = -\frac{1}{N}\int 1 + \ln \rho(\mathbf{r})d\mathbf{r}. \tag{13}$$

In the formalism of DFT, the reactivity parameters are defined as:

$$\text{Chemical potential} = \mu = \left(\frac{\partial E}{\partial N}\right)_{v(\mathbf{r})}. \tag{14}$$

$$\text{Hardness} = \eta = \left(\frac{\partial \mu}{\partial N}\right)_{v(\mathbf{r})} = \left(\frac{\partial^2 E}{\partial N^2}\right)_{v(\mathbf{r})}. \tag{15}$$

$$\text{Softness} = s = \left(\frac{\partial N}{\partial \mu}\right)_{v(\mathbf{r})} = \left(\frac{\partial \rho(\mathbf{r})}{\partial \mu}\right)_{v(\mathbf{r})}. \tag{16}$$

$$\text{Fukui function} = f(\mathbf{r}) = \left(\frac{\partial \rho(\mathbf{r})}{\partial N}\right)_{v(\mathbf{r})}. \tag{17}$$

where μ is the Chemical potential, η is the Hardness, s is the Softness and $f(\mathbf{r})$ is the Fukui Function. Each one of these equations has a specific interpretation in Chemistry. The Chemical potential, μ, is a measure the escaping tendency of an electron, which is constant, through all space, for the ground state of an atom, molecule, or solid [16]. The Hardness, η, is related to the polarizability [15] and the Fukui function [23–25], $\left(\frac{\partial \rho(\mathbf{r})}{\partial N}\right)_{v(\mathbf{r})}$, is interpreted as a measure of the sensitivity of the chemical potential with respect to an external perturbation at a particular point. These properties have been included in the chemical vocabulary since the early 1950s.

2.1. The second derivatives of Shannon's entropy

In this section we obtain the second derivatives of the Shannon entropy respect to the electron density, how a first case, we take the Eq. (11),

$$
\left(\frac{\partial^2 S(\mathbf{r})}{\partial \rho(\mathbf{r}) \partial N} \right)_{v(\mathbf{r})} = \frac{\partial}{\partial \rho(\mathbf{r})} \left\{ -\frac{s(\mathbf{r})}{N^2} - \frac{1}{N} \int \left(\frac{\partial \rho(\mathbf{r})}{\partial N} \right)_{v(\mathbf{r})} \ln \rho(\mathbf{r}) d\mathbf{r} + \frac{1}{N} \right\}
$$

$$
= -\frac{1}{N^2} \int 1 + \ln \rho(\mathbf{r}) d\mathbf{r}
$$

$$
- \frac{1}{N} \int \left(\frac{\partial f(\mathbf{r})}{\partial \rho(\mathbf{r})} \right)_{v(\mathbf{r})} \ln \rho(\mathbf{r}) + \frac{f(\mathbf{r})}{\rho(\mathbf{r})} d\mathbf{r}. \tag{18}
$$

where $f(\mathbf{r})$ is it the Fukui function.

How a second case, consider the Eq. (12),

$$
\left(\frac{\partial^2 S(\mathbf{r})}{\partial \rho(\mathbf{r}) \partial \mu} \right)_{v(\mathbf{r})} = \frac{\partial}{\partial \rho(\mathbf{r})} \left\{ \left[\frac{1}{N} - \frac{s(\mathbf{r})}{N^2} \right] \left(\frac{\partial N}{\partial \mu} \right)_{v(\mathbf{r})} - \int \left(\frac{\partial \rho(\mathbf{r})}{\partial \mu} \right)_{v(\mathbf{r})} \ln \rho(\mathbf{r}) d\mathbf{r} \right\}
$$

$$
= -\frac{1}{N^2} \left(\frac{\partial N}{\partial \mu} \right)_{v(\mathbf{r})} \int 1 + \ln \rho(\mathbf{r}) d\mathbf{r}
$$

$$
- \int \left(\frac{\partial s}{\partial \rho(\mathbf{r})} \right)_{v(\mathbf{r})} \ln \rho(\mathbf{r}) + \frac{s}{\rho(\mathbf{r})} d\mathbf{r}. \tag{19}
$$

where s is the softness.

Finally, for Eq. (13),

$$
\left(\frac{\partial^2 S(\mathbf{r})}{\partial \rho(\mathbf{r})^2} \right)_{v(\mathbf{r})} = -\frac{1}{N} \int \frac{\partial}{\partial \rho(\mathbf{r})} \left(1 + \ln \rho(\mathbf{r}) \right) d\mathbf{r}
$$

$$
= -\frac{1}{N} \int \frac{d\mathbf{r}}{\rho(\mathbf{r})}. \tag{20}
$$

2.2. The change of Shannon entropy respect to the electron energy using Parr-Gadre-Bartolotti model

Now, we obtain the variation of the Shannon entropy respect to the energy, in this case, consider the chemical potential $\mu = \left(\frac{\partial E}{\partial N} \right)_{v(\mathbf{r})}$ and the Eq. (4),

$$\left(\frac{\partial S(\mathbf{r})}{\partial N}\right)_{v(\mathbf{r})}\left(\frac{\partial N}{\partial E}\right)_{v(\mathbf{r})} = -\frac{1}{N}\int\left(\frac{\partial\rho(\mathbf{r})}{\partial N}\right)_{v(\mathbf{r})}\left(\frac{\partial N}{\partial E}\right)_{v(\mathbf{r})}\ln\rho(\mathbf{r})d\mathbf{r}$$

$$+\left[\frac{1}{N}-\frac{s(\mathbf{r})}{N^2}\right]\left(\frac{\partial N}{\partial E}\right)_{v(\mathbf{r})}$$

$$= -\frac{1}{N}\int\left(\frac{\partial\rho(\mathbf{r})}{\partial E}\right)_{v(\mathbf{r})}\ln\rho(\mathbf{r})d\mathbf{r}$$

$$+\left[\frac{1}{N}-\frac{s(\mathbf{r})}{N^2}\right]\left(\frac{\partial N}{\partial E}\right)_{v(\mathbf{r})}, \qquad (21)$$

note the importance of this relation, we can establish a formal relation between the Shannon entropy with the electronic energy[1].

In a similar way, we can obtain the second derivative of the last equation respect to electron density.

$$\left(\frac{\partial^2 S(\mathbf{r})}{\partial\rho(\mathbf{r})\partial E}\right)_{v(\mathbf{r})} = \frac{\partial}{\partial\rho(\mathbf{r})}\left\{-\frac{1}{N}\int\left(\frac{\partial\rho(\mathbf{r})}{\partial E}\right)_{v(\mathbf{r})}\ln\rho(\mathbf{r})d\mathbf{r}+\left[\frac{1}{N}-\frac{s(\mathbf{r})}{N^2}\right]\left(\frac{\partial N}{\partial E}\right)_{v(\mathbf{r})}\right\}$$

$$= -\frac{1}{N}\int\left(\frac{\partial\mu(\mathbf{r})}{\partial\rho(\mathbf{r})}\right)_{v(\mathbf{r})}\ln\rho(\mathbf{r})+\frac{\mu(\mathbf{r})}{\rho(\mathbf{r})}d\mathbf{r}$$

$$-\frac{1}{N^2}\left(\frac{\partial N}{\partial E}\right)_{v(\mathbf{r})}\int 1+\ln\rho(\mathbf{r})d\mathbf{r}. \qquad (22)$$

where $\mu(\mathbf{r})$ it is a local chemical potential.

For obtain a direct application of the last result in the DFT model, we selected the Parr-Gadre-Bartolotti model [26], PGB, this is a local model based in the Thomas-Fermi model [17, 18]. The expression for the energy in the PGB model is,

$$d\,[E]_{PGB} = C\rho(\mathbf{r})^{5/3}+\rho(\mathbf{r})v(\mathbf{r})+BN^{2/3}\rho(\mathbf{r})^{4/3}d\mathbf{r}, \qquad (23)$$

where $B = 0.7544$ and $C = 3.8738$.

Considering Eq. (21), and Eq. (23), the first step is it obtain $\left(\frac{\partial\rho(\mathbf{r})}{\partial E}\right)_{v(\mathbf{r})}$, for do this, consider

$$\left(\frac{\partial\rho(\mathbf{r})}{\partial E_{PGB}}\right)_{v(\mathbf{r})}^{-1} = \left(\frac{\partial E_{PGB}}{\partial\rho(\mathbf{r})}\right)_{v(\mathbf{r})} \qquad (24)$$

---

[1] A more complete study about of the formal relation between electron energy and Shannon entropy will present in other work.

with this and the Eq. (23),

$$\left(\frac{\partial E_{PGB}}{\partial \rho(\mathbf{r})}\right)_{v(\mathbf{r})} = \frac{\partial}{\partial \rho(\mathbf{r})}\left\{C\rho(\mathbf{r})^{5/3} + \rho(\mathbf{r})v(\mathbf{r}) + BN^{2/3}\rho(\mathbf{r})^{4/3}\right\}$$

$$= \frac{5}{3}C\rho(\mathbf{r})^{2/3} + v(\mathbf{r}) + \frac{4}{3}BN^{2/3}\rho(\mathbf{r})^{1/3} \tag{25}$$

and,

$$\left(\frac{\partial \rho(\mathbf{r})}{\partial E_{PGB}}\right)_{v(\mathbf{r})} = \frac{1}{\frac{5}{3}C\rho(\mathbf{r})^{2/3} + v(\mathbf{r}) + \frac{4}{3}BN^{2/3}\rho(\mathbf{r})^{1/3}}, \tag{26}$$

now consider the Eq. (21),

$$\left(\frac{\partial S(\mathbf{r})}{\partial E_{PGB}}\right)_{v(\mathbf{r})} = -\frac{1}{N}\int \frac{\ln \rho(\mathbf{r})}{\frac{5}{3}C\rho(\mathbf{r})^{2/3} + v(\mathbf{r}) + \frac{4}{3}BN^{2/3}\rho(\mathbf{r})^{1/3}}d\mathbf{r} + \left[\frac{1}{N} - \frac{s(\mathbf{r})}{N^2}\right]\left(\frac{\partial N}{\partial E}\right)_{v(\mathbf{r})}. \tag{27}$$

The importance of this results, resides in that for the first time we prove that there exist a formal relationship between concepts of the information theory and the chemical reactivity, based on a formal derivation rather than on a phenomenological interpretation.

2.3. The variation of Shannon's entropy with respect to the electron kinetic energy

In this case, we consider the famous Thomas-Fermi kinetic energy functional, defined as

$$dE_k^{TF}[\rho(\mathbf{r})] = C_F\rho(\mathbf{r})^{5/3}d\mathbf{r}, \qquad C_F = \frac{3}{10}(3\pi^2)^{2/3}, \tag{28}$$

first we use the property (24)

$$\left(\frac{\partial E_k^{TF}}{\partial \rho(\mathbf{r})}\right)_{v(\mathbf{r})} = C_F\frac{5}{3}\rho(\mathbf{r})^{2/3}, \tag{29}$$

now, we use the Eq. (21)

$$\left(\frac{\partial S(\mathbf{r})}{\partial E_k^{TF}}\right)_{v(\mathbf{r})} = -\frac{1}{N}\int \frac{\ln \rho(\mathbf{r})}{C_F\frac{5}{3}\rho(\mathbf{r})^{2/3}}d\mathbf{r} + \left[\frac{1}{N} - \frac{s(\mathbf{r})}{N^2}\right]\frac{1}{\mu},$$

$$= -\frac{3}{5NC_F}\int \frac{\ln \rho(\mathbf{r})}{\rho(\mathbf{r})^{2/3}}d\mathbf{r} - \frac{1}{\mu N^2}\int \rho(\mathbf{r})\ln \rho(\mathbf{r})d\mathbf{r} + \frac{1}{\mu N}. \tag{30}$$

2.4. The variation of Shannon's entropy with respect to the exchange energy in LDA

For simplicity we take only the exchange energy, in the Local Density Approximation (LDA), in this approximation the total energy of a system can be write as the sum of the correlation and exchange energy: $\varepsilon_{cx} = \varepsilon_x + \varepsilon_c$. The correlation part, ε_c has been calculated and the results obtained were expressed like complicated expression of $\rho(\mathbf{r})$ [27]. In our case, we only consider the exchange term, defined as

$$dE_x^{LDA}[\rho(\mathbf{r})] = -\frac{3}{4}\left(\frac{3}{\pi}\right)^{1/3}\rho(\mathbf{r})^{4/3}d\mathbf{r} \tag{31}$$

again,

$$\left(\frac{\partial E_x^{LDA}}{\partial \rho(\mathbf{r})}\right)_{v(\mathbf{r})} - -\left(\frac{3}{\pi}\right)^{1/3}\rho(\mathbf{r})^{1/3}, \tag{32}$$

and

$$\left(\frac{\partial S(\mathbf{r})}{\partial E_x^{LDA}}\right)_{v(\mathbf{r})} = -\frac{1}{N}\int \frac{\ln\rho(\mathbf{r})}{\left(\frac{3}{\pi}\right)^{1/3}\rho(\mathbf{r})^{1/3}}d\mathbf{r} + \left[\frac{1}{N} - \frac{s(\mathbf{r})}{N^2}\right]\left(\frac{\partial N}{\partial E}\right)_{v(\mathbf{r})}$$

$$= -\frac{1}{N}\int \frac{\ln\rho(\mathbf{r})}{\left(\frac{3}{\pi}\right)^{1/3}\rho(\mathbf{r})^{1/3}}d\mathbf{r} - \frac{1}{\mu N^2}\int \rho(\mathbf{r})\ln\rho(\mathbf{r})d\mathbf{r} + \frac{1}{\mu N}. \tag{33}$$

2.5. Variation of Shannon's entropy with respect to the energy, considering the kinetic and exchange effects.

Now, we can take the previous results to obtain the variation on Shannon's entropy considering the kinetic and exchange effects using local models. This derivative will be

$$\left(\frac{\partial S(\mathbf{r})}{\partial E_{k,x}^{TF,LDA}}\right)_{v(\mathbf{r})} = \left(\frac{\partial S(\mathbf{r})}{\partial E_k^{TF}}\right)_{v(\mathbf{r})} + \left(\frac{\partial S(\mathbf{r})}{\partial E_x^{LDA}}\right)_{v(\mathbf{r})}$$

$$= -\frac{3}{5NC_F}\int \frac{\ln\rho(\mathbf{r})}{\rho(\mathbf{r})^{2/3}}d\mathbf{r} - \frac{1}{\mu N^2}\int \rho(\mathbf{r})\ln\rho(\mathbf{r})d\mathbf{r} + \frac{1}{\mu N}$$

$$-\frac{1}{N}\int \frac{\ln\rho(\mathbf{r})}{\left(\frac{3}{\pi}\right)^{1/3}\rho(\mathbf{r})^{1/3}}d\mathbf{r} - \frac{1}{\mu N^2}\int \rho(\mathbf{r})\ln\rho(\mathbf{r})d\mathbf{r} + \frac{1}{\mu N}$$

$$= \left[-\frac{3}{5NC_F}\int \frac{\ln\rho(\mathbf{r})}{\rho(\mathbf{r})^{2/3}}d\mathbf{r} - \frac{1}{N}\int \frac{\ln\rho(\mathbf{r})}{\left(\frac{3}{\pi}\right)^{1/3}\rho(\mathbf{r})^{1/3}}d\mathbf{r}\right]$$

$$-\frac{1}{\mu N^2} \int \rho(\mathbf{r}) \ln \rho(\mathbf{r}) d\mathbf{r} + \frac{1}{\mu N}$$

$$= -\frac{1}{N} \left(\frac{3}{5C_F} + \frac{1}{(3/\pi)^{1/3}} \right) \int \left(\frac{1+\rho(\mathbf{r})^{1/3}}{\rho(\mathbf{r})^{2/3}} \right) \ln \rho(\mathbf{r}) d\mathbf{r}$$

$$-\frac{1}{\mu N^2} \int \rho(\mathbf{r}) \ln \rho(\mathbf{r}) d\mathbf{r} + \frac{1}{\mu N}$$

$$= -\int \left[A \left(\frac{1+\rho(\mathbf{r})^{1/3}}{\rho(\mathbf{r})^{2/3}} \right) - B\rho(\mathbf{r}) \right] \ln \rho(\mathbf{r}) d\mathbf{r} + BN, \qquad (34)$$

where

$$A = \frac{3}{5NC_F} + \frac{1}{N(3/\pi)^{2/3}},$$

and

$$B = \frac{1}{\mu N^2}.$$

2.6. Summary of relationships obtained

Finally, we present a summary of the different relations obtained in this work. Whit this results, is possible say that the information theory is a model that is subjacent to the Density Functional Theory.

In the follow section we show some results of this relations applied to some chemical process.

$$\left(\frac{\partial S(\mathbf{r})}{\partial N} \right)_{v(\mathbf{r})} = -\frac{s(\mathbf{r})}{N^2} - \frac{1}{N} \int \left(\frac{\partial \rho(\mathbf{r})}{\partial N} \right)_{v(\mathbf{r})} \ln \rho(\mathbf{r}) d\mathbf{r} + \frac{1}{N}. \qquad (35)$$

$$\left(\frac{\partial S(\mathbf{r})}{\partial \mu} \right)_{v(\mathbf{r})} = \left[\frac{1}{N} - \frac{s(\mathbf{r})}{N^2} \right] \left(\frac{\partial N}{\partial \mu} \right)_{v(\mathbf{r})} - \frac{1}{N} \int \left(\frac{\partial \rho(\mathbf{r})}{\partial \mu} \right)_{v(\mathbf{r})} \ln \rho(\mathbf{r}) d\mathbf{r}. \qquad (36)$$

$$\left(\frac{\partial S(\mathbf{r})}{\partial \rho(\mathbf{r})} \right)_{v(\mathbf{r})} = -\frac{1}{N} \int 1 + \ln \rho(\mathbf{r}) d\mathbf{r}. \qquad (37)$$

$$\left(\frac{\partial S(\mathbf{r})}{\partial E} \right)_{v(\mathbf{r})} = -\frac{1}{N} \int \left(\frac{\partial \rho(\mathbf{r})}{\partial E} \right)_{v(\mathbf{r})} \ln \rho(\mathbf{r}) d\mathbf{r} + \left[\frac{1}{N} - \frac{s(\mathbf{r})}{N^2} \right] \left(\frac{\partial N}{\partial E} \right)_{v(\mathbf{r})}. \qquad (38)$$

$$\left(\frac{\partial^2 S(\mathbf{r})}{\partial \rho(\mathbf{r}) \partial N} \right)_{v(\mathbf{r})} = -\frac{1}{N} \int \left(\frac{\partial f(\mathbf{r})}{\partial \rho(\mathbf{r})} \right)_{v(\mathbf{r})} \ln \rho(\mathbf{r}) + \frac{f(\mathbf{r})}{\rho(\mathbf{r})} d\mathbf{r}. \qquad (39)$$

$$\left(\frac{\partial^2 S(\mathbf{r})}{\partial \rho(\mathbf{r}) \partial \mu} \right)_{v(\mathbf{r})} = -\int \left(\frac{\partial s}{\partial \rho(\mathbf{r})} \right)_{v(\mathbf{r})} \ln \rho(\mathbf{r}) + \frac{s}{\rho(\mathbf{r})} d\mathbf{r}. \qquad (40)$$

$$\left(\frac{\partial^2 S(\mathbf{r})}{\partial \rho(\mathbf{r})^2} \right)_{v(\mathbf{r})} = -\frac{1}{N} \int \frac{d\mathbf{r}}{\rho(\mathbf{r})}. \qquad (41)$$

$$\left(\frac{\partial^2 S(\mathbf{r})}{\partial \rho(\mathbf{r})\partial E}\right)_{v(\mathbf{r})} = -\frac{1}{N^2}\left(\frac{\partial N}{\partial E}\right)_{v(\mathbf{r})}\int 1 + \ln\rho(\mathbf{r})d\mathbf{r}. \tag{42}$$

$$\left(\frac{\partial S(\mathbf{r})}{\partial E_k^{TF}}\right)_{v(\mathbf{r})} = -\frac{3}{5NC_F}\int\frac{\ln\rho(\mathbf{r})}{\rho(\mathbf{r})^{2/3}}d\mathbf{r} - \frac{1}{\mu N^2}\int\rho(\mathbf{r})\ln\rho(\mathbf{r})d\mathbf{r} + \frac{1}{\mu N}. \tag{43}$$

$$\left(\frac{\partial S(\mathbf{r})}{\partial E_x^{LDA}}\right)_{v(\mathbf{r})} = -\frac{1}{N}\int\frac{\ln\rho(\mathbf{r})}{\left(\frac{3}{\pi}\right)^{1/3}\rho(\mathbf{r})^{1/3}}d\mathbf{r} - \frac{1}{\mu N^2}\int\rho(\mathbf{r})\ln\rho(\mathbf{r})d\mathbf{r} + \frac{1}{\mu N}. \tag{44}$$

$$\left(\frac{S(\mathbf{r})}{E_{k,x}^{TF,LDA}}\right)_{v(\mathbf{r})} = -\int\left[A\left(\frac{1+\rho(\mathbf{r})^{1/3}}{\rho(\mathbf{r})^{2/3}}\right) - B\rho(\mathbf{r})\right]\ln\rho(\mathbf{r})d\mathbf{r} + BN. \tag{45}$$

$$\left(\frac{\partial S(\mathbf{r})}{\partial E^{PGB}}\right)_{v(\mathbf{r})} = -\frac{1}{N}\int\frac{\ln\rho(\mathbf{r})}{\frac{5}{3}C\rho(\mathbf{r})^{2/3} + v(\mathbf{r}) + \frac{4}{3}BN^{2/3}\rho(\mathbf{r})^{1/3}}d\mathbf{r}$$
$$+ \left[\frac{1}{N} - \frac{s(\mathbf{r})}{N^2}\right]\left(\frac{\partial N}{\partial E}\right)_{v(\mathbf{r})}. \tag{46}$$

3. Description of a simple chemical process

3.1. Reaction $CH_2CHF + CH_3 \rightarrow CH_3CHFCH_2$

To show the application of the relations obtained, we have selected the following radical-molecule chemical reaction: $CH_2CHF + CH_3 \rightarrow CH_3CHFCH_2$.

Structures and energies have been obtained along the reaction path using m062x/6-311++G(d,p) density functional method with Gaussian 09 [31]. The electron density was calculated with Pérez-Jordá's algorithms [8] and a D-Grid 4.6 [9]. Molecular Electrostatic Potential (MEP) isosurfaces were obtained with Molden 5.0 [22]. The Fukui function condensed was calculated using natural atomics orbitals obtained in a Natural Population Analysis [28].

The condensed Fukui function were calculated according to the following approximations:

$$f(\mathbf{r})^+ = |\phi(\mathbf{r})_{LUMO}|^2 + \sum_{i=1}^{n}\frac{\partial}{\partial N}|\phi(\mathbf{r})|^2. \tag{47}$$

$$f(\mathbf{r})^- = |\phi(\mathbf{r})_{HOMO}|^2 + \sum_{i=1}^{n}\frac{\partial}{\partial N}|\phi(\mathbf{r})|^2. \tag{48}$$

$$f(\mathbf{r}) = \frac{1}{2}\left(f(\mathbf{r})^+ + f(\mathbf{r})^-\right). \tag{49}$$

where $\phi(\mathbf{r})_{LUMO}$ correspond to LUMO electron density and $\phi(\mathbf{r})_{HOMO}$ to HOMO electron density. When a molecule accepts electrons, the electrons tend to go to places where $f(\mathbf{r})^+$ is large because it is at these locations that the molecule is most able to stabilize additional electrons. Therefore a molecule is susceptible of a nucleophilic attack at sites where $f(\mathbf{r})^+$

is large. Similarly, a molecule is susceptible of an electrophilic attack at sites where $f(\mathbf{r})^-$ is large, because these are the regions where electron removal destabilizes the molecule the least. In chemical density functional theory, the Fukui functions are the key regioselectivity indicators for electron-transfer controlled reactions. In order to use these expressions we have chosen the Natural Population Analysis, NPA, which involves only matrix diagonalization of small subsets of the density matrix, and also requires a negligible amount of computer time. Although it is more involved than a Mulliken or Löwdin analysis, for a theoretical analysis using von Neumann entropies, NPA is an attractive method [29, 30, 34].

In Figure (1(a)) we present the electron energy profile of the reaction as a function of the reaction coordinate RX in Å. Trends in the condensed Fukui function are shown in Figure (1(b)). Analogously, trends in Hardness and chemical potential are shown in Figures (1(c)) and (1(d)), respectively.

In Figures (2(a)) and (2(b)) we present the trends in the kinetic and exchange energies as a function of the reaction coordinate.

Energy values show that the reaction is exothermic, with the following energy values at stationary points: E(reactants)=-217.629652 A.U.; E(transition state)= -217.615808 A.U. and E(products)= -217.674564 A.U. One of the main points of interest for our purposes is the analysis of the structures along the reaction path in terms of descriptors that are related to chemical reactivity. In this specific case, we consider the kinetic and exchange energies (see Figs. (2(a)) and (2(b))). For the kinetic energy we note that there exists a region limited by a maximum at $RX = -0.64230$ and a minimum at $RX = 1.44550$, in which important chemical changes occurs. It is associated with bond forming, in the Figures (2(c))-(2(l)) present the Molecular Electrostatic Potential of this process, probably this result permit us establish a kinetically classification criteria of this reaction, that is, the principal parameter that govern this process it is the kinetic. The minima in $RX = 1.44550$ is it in relation to important chemical changes that occurs in the frontier orbitals, see the Figures (2(m)), (2(n)) (2(o)) and (2(p)) where present the isosurfaces of the HOMO and LUMO orbitals.

In the case of the Hardness, we note that not are a correspondence between the maximum or minima of the electron, kinetic or exchange energy with the minimum and maximum of the Hardness. This points, again are related at with the structural changes that occurs in this zone, also, of course, this changes involved important changes in the frontier orbitals. A similar aspect occurs with the condensed Fukui functions, Figure (1(b)), where we note that in the point $RX = 0.32119$ exist a equality $f(\mathbf{r})^+ \simeq f(\mathbf{r})^- \simeq f(\mathbf{r}) \simeq 0.5$ with this numerical result we can note a parallelism with the relation of a hard/soft acids/basis, proposed by Pearson. That in terms of the condensed Fukui functions would be, when a chemical process occurs, exist a point where the active sites of the structures have a equalization of a kind of chemical attack, nucleophilic or electrophilic. See, the Figures (3(a)), (3(b)) and (3(c)).

In this sense, is important to note that, in this case we suspect that the more important changes in the parameters and their equalization occurs in the transition state, this can be suspect by chemical intuition, but how we can see in the different graphics this not occurs, in fact, the hardness, Fig. (1(c)), that exist a minima in $RX = -0.64230$ and a maximum in $RX = 3.21128$ this zone permit us define as a zone where the process occurs via nucleophilic attack.

In Figure (3(e)) and (3(d)) we present, the tendencies of Shannon entropy and the electron energy with the LDA approximation, E^{LDA}. In this case, the E^{LDA} increase basically in a linear way, but in the point $RX = 2.73035$ have a maximum, this maximum appears to in the chemical potential tendency. Respect to the Shannon entropy we note more structure than E^{LDA}, and in this case, we note than the Shannon entropy have a similar tendency than the electron energy, in the same form than the energy the Shannon entropy can detect the transition state, see the Figure (3(f)).

By comparison between the Shannon entropy and electron energy profiles we can note that the Shannon entropy, have a zone delimited in the region $-0.5 \leq RX \leq 0.5$, see Figure (3(f)), that have a correspondence with the transition state zone, this is an important observation, because, the Shannon entropy can detect a transition state zone, where occurs a transfer and redistribution of electron density. Comparing, the tendencies of condensed Fukui function, with the Shannon entropy tendency, we note that this zone, have a relation with a charge transfer between the methyl and the molecule and is in this zone where occurs the process of bond forming. In the same way that the condensed Fukui function, this zone is where the subtle interactions among frontier orbitals occurs, and permit to the molecular system start a complex process of the chemical bond forming, in Figure (4(a)) we present the tendency of a normal mode of vibration, and we note that, the zone predicted by the Shannon entropy exhibit a correspondence with a zone of the negative values of the frequencies, consequently, is possible that we can not speak of a specific point in the reaction path where occurs a bond breaking or bond forming. A similar argumentation, can be applied to the description of the transition state zone. Note the importance of this, is possible that, still with the modern techniques of the experimental chemistry, such the femtochemistry [19–21], we can not detect just a transition state structure, or else, a zone of transition. This zone of transition, is accoted by a zone where the Shannon entropy tendency have a slope approximately to zero, see the Figure (4(b)) accordingly, we can say that the Shannon entropy can detect and predict the zone where the most important chemical changes will be occurs, so, this kind of entropy permit us reveals some chemical aspects that are subjacent in a chemical process. Other important, observation is that the maximum of the electron energy, not correspond to the minimum of the frequencies, it is represents, probability, a conflict between the convectional interpretation, in the sense that the maximum of the energy correspond at one possible transition state structure that, in general, have the most negative value of the frequencies, but how we show in this case, this not occurs, by other hand, the minimum value of the frequencies it is related at one equality with the condensed Fukui functions, see the Figure (4(c)).

In general terms, is important to note, that the frequencies tendencies have a tree zones of negative frequencies, these Hessian values represent the transition vector which show maxima at the vicinity of the transition state. Several features are worth mentioning, the TS corresponds indeed to a saddle point, maxima at the Hessian correspond to high kinetic energy values (largest frequencies for the energy cleavage reservoirs) since they fit with maximal values in the entropy profile, and the Hessian is minimal at the TS, where the kinetic energy is the lowest (minimal molecular frequency) and it corresponds to a saddle point. In this case, the analysis of frequencies can give us, a general idea about that this mechanics occurs in tree steps, in each one, occurs some structural rearrays at expense of a decrease in energy exchange and increased kinetic energy. It is relevant note, that the Shannon entropy, see Figure (4(b)), can detect three zones where the frequencies are negative, the first of them is between $-3.5 \leq RX \leq -2.2$ the second in $-0.5 \leq RX0.5$ and the last in $3.8 \leq RX \leq 4.0$. In the first and third zone, the Shannon entropy exhibits a change in their curvature, also, note that this last observation have a correspondence with the other parameters such, hardness, softness, chemical potential and with some Shannon entropy derivatives.

In Figure (5(b)) we compare the trend of the derivative of Snannon's entropy with respect to the number of electrons: $\left(\frac{\partial S(\mathbf{r})}{\partial N}\right)_{v(\mathbf{r})}$ as a function of RX, with that of the DFT chemical reactivity descriptor: $\left(\frac{\partial \mu}{\partial N}\right)_{v(\mathbf{r})}$, which is associated to hardness. Even though absolute values are not the same everywhere, there is a perfect coincidence in the region close to the transition state, in terms of the RX position as well as in the absolute values.

A similar situation occurs with the derivative of Shannon's entropy with respect to the chemical potential: $\left(\frac{\partial S(\mathbf{r})}{\partial \mu}\right)_{v(\mathbf{r})}$, and the inverse of the Exchange energy,

$$\frac{1}{E_i^{LDA}} = -\frac{1}{\left(\frac{3}{\pi}\right)^{1/3} \int \rho(\mathbf{r})^{4/3} d\mathbf{r}},$$

see Figure (5(d)).

Finally, a plot of the derivative of Shannon's entropy with respect to the electronic energy $\left(\frac{\partial S(\mathbf{r})}{\partial E}\right)_{v(\mathbf{r})}$ as a function of RX, behaves in a manner that is remarkably similar to that of the DFT descriptor for softness: $\left(\frac{\partial N}{\partial \mu}\right)_{v(\mathbf{r})}$ (Figure (5(f)).

The relations that were suggested above based on numerical evidence can be summarized as:

$$\left(\frac{\partial S(\mathbf{r})}{\partial N}\right)_{v(\mathbf{r})} \simeq \left(\frac{\partial \mu}{\partial N}\right)_{v(\mathbf{r})}. \tag{50}$$

$$\left(\frac{\partial S(\mathbf{r})}{\partial \mu}\right)_{v(\mathbf{r})} \simeq -\frac{3}{4}\left(\frac{3}{\pi}\right)^{1/3} \int \rho(\mathbf{r})^{4/3} d\mathbf{r}. \tag{51}$$

$$\left(\frac{\partial S(\mathbf{r})}{\partial E}\right)_{v(\mathbf{r})} \simeq \left(\frac{\partial N}{\partial \mu}\right)_{v(\mathbf{r})}. \tag{52}$$

From the numerical results obtained from Eqs. (50) and (52), we have been able to establish a linear relationship between the parameters involved, and the following expressions have been obtained:

$$\text{Hardness} = \left(\frac{\partial \mu}{\partial N}\right)_{v(\mathbf{r})} = 28.3141\left(\frac{\partial S(\mathbf{r})}{\partial N}\right)_{v(\mathbf{r})} + 0.0226, \tag{53}$$

$$\text{Softness} = \left(\frac{\partial N}{\partial \mu}\right)_{v(\mathbf{r})} = 24.1088\left(\frac{\partial S(\mathbf{r})}{\partial E}\right)_{v(\mathbf{r})} + 0.5876, \tag{54}$$

Thus, the hardness and softness values of this chemical reaction can be obtained with a good level of accuracy from the derivatives of Shannon's entropy.

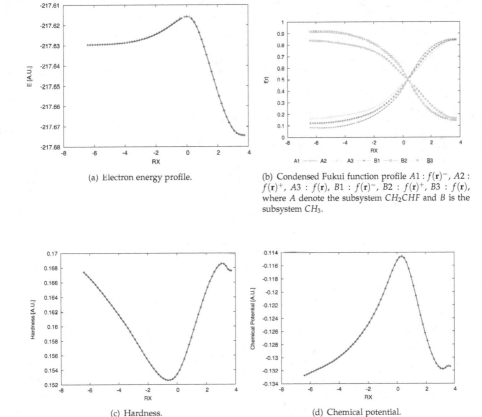

(a) Electron energy profile.

(b) Condensed Fukui function profile $A1 : f(\mathbf{r})^-$, $A2 : f(\mathbf{r})^+$, $A3 : f(\mathbf{r})$, $B1 : f(\mathbf{r})^-$, $B2 : f(\mathbf{r})^+$, $B3 : f(\mathbf{r})$, where A denote the subsystem CH_2CHF and B is the subsystem CH_3.

(c) Hardness.

(d) Chemical potential.

Figure 1. Trends of the reaction $CH_2CHF + CH_3 \rightarrow CH_3CHFCH_2$.

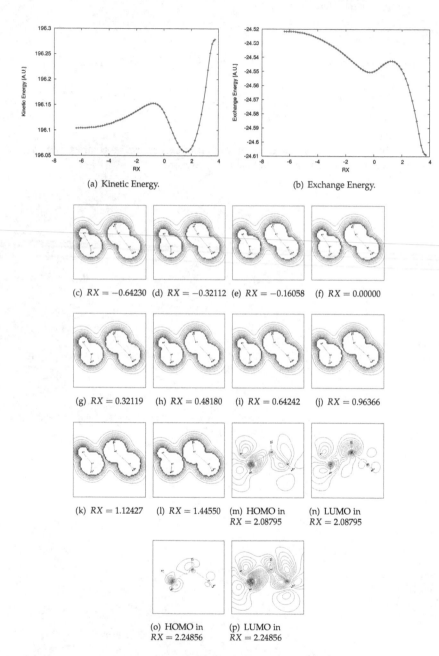

(a) Kinetic Energy.   (b) Exchange Energy.

(c) $RX = -0.64230$ (d) $RX = -0.32112$ (e) $RX = -0.16058$ (f) $RX = 0.00000$

(g) $RX = 0.32119$ (h) $RX = 0.48180$ (i) $RX = 0.64242$ (j) $RX = 0.96366$

(k) $RX = 1.12427$ (l) $RX = 1.44550$ (m) HOMO in   (n) LUMO in
                                          $RX = 2.08795$ $RX = 2.08795$

(o) HOMO in   (p) LUMO in
$RX = 2.24856$ $RX = 2.24856$

Figure 2. Isosurfaces of the Molecular Electrostatic Potential, in $-0.64230 \leq RX \leq 1.44550$.

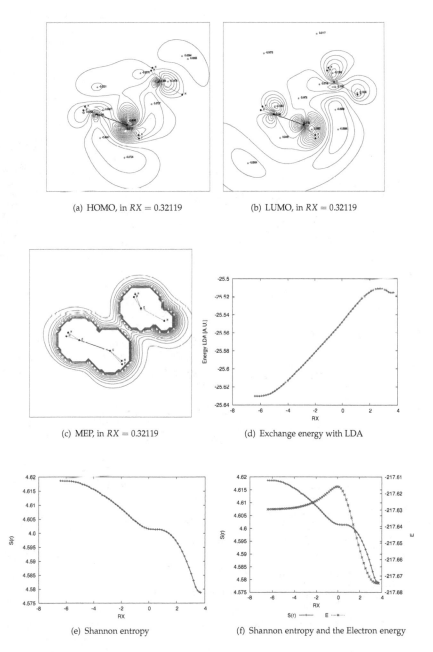

(a) HOMO, in $RX = 0.32119$

(b) LUMO, in $RX = 0.32119$

(c) MEP, in $RX = 0.32119$

(d) Exchange energy with LDA

(e) Shannon entropy

(f) Shannon entropy and the Electron energy

Figure 3. Trends of the reaction $CH_2CHF + CH_3 \rightarrow CH_3CHFCH_2$.

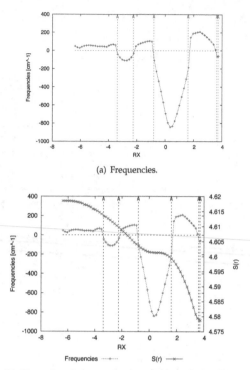

(a) Frequencies.

(b) Comparison between the frequencies and the Shannon entropy.

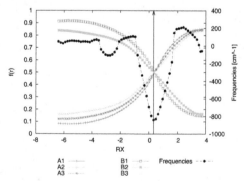

(c) Comparison between the condensed Fukui function and the Frequencies.
Where $A1 : f(\mathbf{r})^-$, $A2 : f(\mathbf{r})^+$, $A3 : f(\mathbf{r})$, $B1 : f(\mathbf{r})^-$, $B2 : f(\mathbf{r})^+$, $B3 : f(\mathbf{r})$, where A denote the subsystem CH_2CHF and B is the subsystem CH_3

Figure 4. Comparison among Frequencies, Shannon entropy and Fukui function.

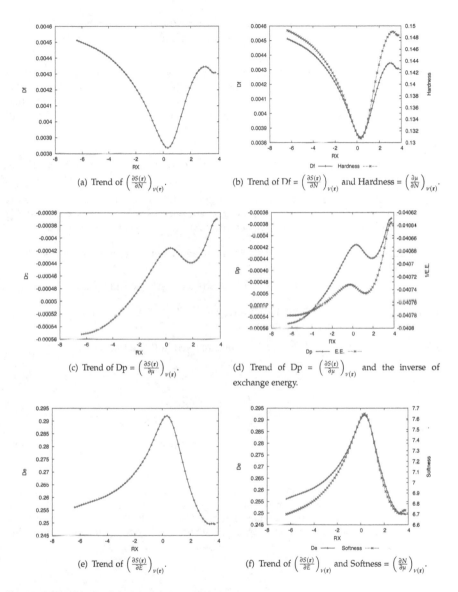

Figure 5. Trends of the first derivatives of Shannon entropy.

3.2. Ethane conformational analysis

In this section we present a conformational analysis of ethane, as a function of the dihedral rotation angle between the two methyl groups. The initial point corresponds to the eclipsed conformer, and an Intrinsic Reaction Coordinate analysis (IRC) has been performed, with ten degrees steps in the dihedral angle. This system was calculated in Gaussian 03 [32], using B3LYP/cc-pVTZ.

In Figure (6(a)) we present the electron energy and Shannon entropy profiles. An excellent agreement is observed between the two curves. Furthermore, a clear similarity is also observed between the behavior, along the reaction path, of the hardness and of the first derivative of Shannon's entropy with respect to the number of electrons. This is shown in Figure (6(b)).

In Figure (6(c)) we show the kinetic and exchange energies as a function of RX. Both quantities exhibit inverse behaviors,the position of the minima in one of them coinciding with the maxima of the other.

One of the most important results is shown in Figure (6(d)), where the exchange energy is plotted against Shannon's entropy along the reaction path. A perfect linear correlation is observed between these apparently unconnected quantities, leading to conclude that

i) The Shannon entropy in position space can be used as a measure of the exchange effects in molecular systems.

ii) There exists a direct relationship between the Shannon entropy and the exchange energy.

Based in these conclusions and results, a new research line is being developed, to construct a functional based on an entropic criterion.

The numerical results obtained above cab be summarized in the following manner:

$$\left(\frac{\partial S(\mathbf{r})}{\partial N}\right)_{v(\mathbf{r})} \simeq \left(\frac{\partial N}{\partial \mu}\right)_{v(\mathbf{r})}. \tag{55}$$

$$S(\mathbf{r}) \simeq -\frac{3}{4}\left(\frac{3}{\pi}\right)^{1/3}\int \rho(\mathbf{r})^{4/3}d\mathbf{r}. \tag{56}$$

The last equation suggests that there may be a fundamental connection between Chemical Density Functional Theory and Information Theory. In Table (1), we present the numerical results of Shannon's entropy and exchange energy. Using these results and a least square regression, is it possible to obtain the following linear relation between Shannon entropy and exchange energy:

$$E_e = AS(\mathbf{r}) - B, \tag{57}$$

where $A = 2.2127$ and $B = 21.0061$. With this equation, we can reproduce the Exchange Energy with a precision of 1×10^{-6}.

Angle	$S(\mathbf{r})$ [nats]	E.E. [A.U.]	$E_e = AS(\mathbf{r}) - B$ [A.U.]	precision
0	4.5902766	-10.849190	-10.8492232214	0.9999
10	4.5901580	-10.849449	-10.8494856465	0.9999
20	4.5898836	-10.850050	-10.8500928087	0.9999
30	4.5895363	-10.850815	-10.8508612761	0.9999
40	4.5892201	-10.851516	-10.8515609289	0.9999
50	4.5890276	-10.851953	-10.8519868718	0.9999
60	4.5889557	-10.852113	-10.8521459642	0.9999
70	4.5890276	-10.851953	-10.8519868718	0.9999
80	4.5892201	-10.851516	-10.8515609289	0.9999
90	4.5895363	-10.850815	-10.8508612761	0.9999
100	4.5898836	-10.850050	-10.8500928087	0.9999
110	4.5901580	-10.849449	-10.8494856465	0.9999
120	4.5902766	-10.849190	-10.8492232214	0.9999
130	4.5901580	-10.849449	-10.8494856465	0.9999
140	4.5898836	-10.850050	-10.8500928087	0.9999
150	4.5895363	-10.850815	-10.8508612761	0.9999
160	4.5892201	-10.851516	-10.8515609289	0.9999
170	4.5890276	-10.851953	-10.8519868718	0.9999
180	4.5889557	-10.852113	-10.8521459642	0.9999

Table 1. Numerical values of Shannon's entropy and Exchange Energy, in the ethane conformational analysis, where $A = 2.2127$ and $B = 21.0061$.

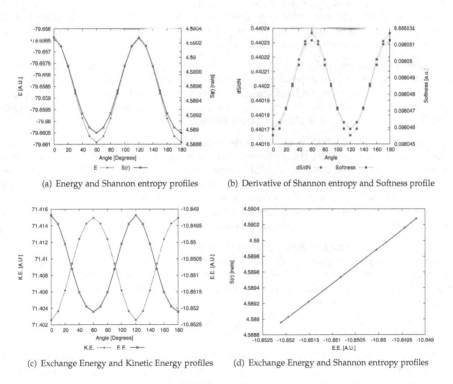

(a) Energy and Shannon entropy profiles

(b) Derivative of Shannon entropy and Softness profile

(c) Exchange Energy and Kinetic Energy profiles

(d) Exchange Energy and Shannon entropy profiles

Figure 6. Tendencies obtained for the Ethane Conformational Analysis.

4. Conclusion

In this work, I have derived relationships that connect Shannon's entropy and its derivatives, with well-known concepts in Density Functional Theory. Numerical applications of these relationships have been performed for two simple problems.

This has been permit us start a new investigation line about it, and with some this results we continue whit the study of the formalism for the construction of a functional based in a principles of physics, and the information theory, also, we pretend continue the develop of the some models that permit us find the direct relation between electron energy with the Shannon entropy.

By other hand, the application of the concepts of the information theory permit do a description more accurate than the description based in only a energetically criteria, and continue with the spirit of some works of Frieden, we speculate that is possible define or found a form that derived the DFT from some fundamental expression, that come form the Information Theory, as Frieden derivations of the fundamental equations of the Thermodynamics, or the derivation of the Scrhödinger equation.

Finally, with this example we have tried to link information from a system that is subjected to a process with the physical and chemical changes. Thus, we have linked the concept of *information*, which is an epistemological concept completely with ontological concepts and the solution concepts or interpretation of the results allows us feedback on these concepts in ontological terms, according to the author, abstract and more general.

By other part, is probable that today do not exist a ortodoxical definition of what actually is the *information*, beyond that presented by Shannon and its guidelines, criteria, characterization of it, among other things, the interpretation and the relationship with other concepts such as: energy, electron density, chemical reactivity parameters and many others need be discussed to try of establish a formal relation between concepts.

So, there is no doubt that both knowledge and the praxis and reality as knowledge scientific understanding and also, is it clear that information concept and the model itself is interdisciplinary or transdisciplinary. The concept and moreover, the model itself, promotes a systematic relation with causal analogies and parallelism with scientific knowledge, which transcends the framework of the source domain and extend in various directions, thus making the knowledge acquires an unusual resonance, as this, we believe it is feasible to complement the explanations of natural processes and natural systems.

This model is not intended that the manner of the old school, that using metaphysical substance, the particularities of the processes reveal themselves to us in the end as a progressive manifestation of homogeneous order or a unitary whole and absolute. It is simply to promote and implement a partnership scheme which promises analog route and cover knowledge in a way easier.

Acknowledgments

I wish to thank José María Pérez Jordá and M. Kohout for kindly providing their numerical codes. I acknowledge financial support from Prof. Annik Vivier-Bunge through project "Red de Química Teórica para el Medio Ambiente y Salud" and for helpful discussions.

Author details

Nelson Flores-Gallegos

Universidad Autónoma Metropolitana-Iztapalapa, México

References

[1] B. Roy Frieden. Fisher information as the basis for the Schrödinger wave equation. *Am. J. Phys.* 1989, *11*, 1004-1008.

[2] B. Roy Frieden. Fisher information, disorder, and the equilibrium distributions of physics. *Phys. Rev. A* 1990, *41*, 4265-4276.

[3] B. Roy Frieden and Roy J. Hughes. Spectral $1/f$ noise derived from extremized physical information. *Phys. Rev. E* 1994, *49*, 2644-2649.

[4] B. Roy Frieden and Bernard H. Soffer. Lagrangians of physics and the game of Fisher-information transfer. *Phys. Rev. E* 1995, *52*, 2274-2286.

[5] B. Roy Frieden and W. J. Cocke. Foundation for Fisher-information-based derivations of physical laws *Phys. Rev. E* 1996, *54*, 257-260.

[6] Parveen Fazal, S., Sen, K.D., Gutierrez G., and Fuentealba, P. Shannon entropy of 1-normalized electron density. *Indian Journal of Chemistry.* 2000, *39A*, 48-49.

[7] Shannon, C. E. A Mathematical Theory of Communication. *The Bell System Thecnical Journal.* 1948, *27*, 379-423.

[8] Pérez-Jordá, Jose M., Becke, Axel D. and San-Fabian, Emilio. Automatic numerical integration techniques for polyatomic molecules. *J. Chem. Phys.* 1994, *100*, 6520-6534.

[9] M. Kohout. DGrid, version 4.6, Radebeul, 2011.

[10] Nelson Flores-Gallegos and Rodolfo O. Esquivel. von Neumann Entropies Analysis in Hilbert Space for the Dissociation Processes of Homonuclear and Heteronuclear Diatomic Molecules *J. Mex. Chem. Soc.*, 2008, *52*, 19-30.

[11] Rodolfo O. Esquivel, Nelson Flores-Gallegos, Cristina Iuga, Edmundo Carrera, Juan Carlos Angulo and Juan Antolín. Phenomenological description of the transition state, and the bond breaking and bond forming processes of selected elementary chemical reactions: An information-theoretic study. *Theoretical Chemistry Accounts.*, 2009, *124*, 445-460.

[12] R.O. Esquivel, N. Flores-Gallegos, C. Iuga, E. Carrera, J.C. Angulo and J. Antolín. Phenomenological description of selected elementary chemical reaction mechanisms: An information-theoretic study. *Physics Letters A, 2010* , *374*, 948-951.

[13] Rodolfo O. Esquivel, Juan Carlos Angulo, Juan Antolín, Jesús S. Dehesa, Sheila López-Rosa and Nelson Flores-Gallegos. Complexity analysis of selected molecules

in position and momentum spaces. *Physical Chemistry Chemical Physics*, 2010 ,   12, 7108-7116.

[14] Gadre, S.R., Sears, S.B. Some novel characteristics of atomic information entropies. *Phys. Rev. A.*, *1985*, 32, 2602-2606.

[15] R. G. Pearson. Hard and Sotf acids and bases. *J. Am. Chem. Soc.*, *1963*, 85, 3533-3539.

[16] R. G. Parr and W. Yang. Density functional approach to the frontier electron theory of the chemical reactivity. *J. Am. Chem. Soc.*, *1984*, 106, 4049-4050.

[17] L. H. Thomas. The calculation of atomic fields. *Proc. Camb. Phil. Soc.*, *1927*, 23, 542-548.

[18] E. Fermi. Un metodo statistce per la determinazione di algune propieta dell'atomo. *Rend. Accad., Licei*, *1927*, 54, 2627-2630.

[19] A. H. Zawail. Laser Femtochemistry. *Science*, *1988*, 242, 1645-1653.

[20] A. H. Zewail. Femtochemistry: Atomic-Scale Dynamics of the Chemical Bond. *J Phys. Chem. A*, *2000*, 104, 5660-5694.

[21] J. S. Baskin and A. H. Zewail. Freezing Time - in a Femtosecond. *Sci. Spectra*, *1998*, *14*, 62-71.

[22] Schaftenaar, G. and Noordik, J. H., MOLDEN: a pre- and post-processing program for molecular and electronic structures. *J. Comput. Aided Mol. Design.*, *2000*, 123-134.

[23] K. Fukui. The role of frontier orbitals in chemical reactions. *Science*, *1987*, 218, 747-754.

[24] K. Fukui, T. Yonezawa and H. Shingu. A molecular orbital theory of reactivity in aromatic hydrocarbons. *J. Chem. Phys.*, *1952*, 20, 722-725.

[25] K. Fukui, T. Yonezawa and C. Nagata. Molecular orbital theory of orientation in aromatic heteroaromatic, and other conjugated molecules. *J. Chem. Phys.*, *1952*, 22, 1433-1442.

[26] R. G. Parr, S. R. Gadre and L. J. Bartolotti. Local density functional theory of atoms and molecules. *Proc. Natl. Acad. Sci. USA*, *1979*, 76, 2522-2526.

[27] S. H. Vosko, L. Wilk and M. Nusair. Accurate spin-dependent electron liquid correlation energies for local spin density calculations: a critical analysis. *Can. J. Phys.*, *1980*, 58, 1200-1211.

[28] Reed A. E., Weinstock R. B., Weinhold F. Natural Population Analysis. *J. Chem. Phys.* Vol. 83, No. 2, (1985) 735-747.

[29] Nelson Flores-Gallegos *Teoría de información cuántica como lenguaje conceptual en Química*. Tesis Doctoral. Universidad Autónoma Metropolitana-Iztapalapa. México D.F. (2010).

[30] Edmundo M. Carrera, Nelson Flores-Gallegos, Rodolfo O. Esquivel. Natural atomic probabilities in quantum information theory. *Journal of Computational and Applied Mathematics.*, Vol. 233, 1483-1490 (2010).

[31] Gaussian 09, Revision B.01, M. J. Frisch, G. W. Trucks, H. B. Schlegel, G. E. Scuseria, M. A. Robb, J. R. Cheeseman, G. Scalmani, V. Barone, B. Mennucci, G. A. Petersson, H. Nakatsuji, M. Caricato, X. Li, H. P. Hratchian, A. F. Izmaylov, J. Bloino, G. Zheng, J. L. Sonnenberg, M. Hada, M. Ehara, K. Toyota, R. Fukuda, J. Hasegawa, M. Ishida, T. Nakajima, Y. Honda, O. Kitao, H. Nakai, T. Vreven, J. A. Montgomery, Jr., J. E. Peralta, F. Ogliaro, M. Bearpark, J. J. Heyd, E. Brothers, K. N. Kudin, V. N. Staroverov, T. Keith, R. Kobayashi, J. Normand, K. Raghavachari, A. Rendell, J. C. Burant, S. S. Iyengar, J. Tomasi, M. Cossi, N. Rega, J. M. Millam, M. Klene, J. E. Knox, J. B. Cross, V. Bakken, C. Adamo, J. Jaramillo, R. Gomperts, R. E. Stratmann, O. Yazyev, A. J. Austin, R. Cammi, C. Pomelli, J. W. Ochterski, R. L. Martin, K. Morokuma, V. G. Zakrzewski, G. A. Voth, P. Salvador, J. J. Dannenberg, S. Dapprich, A. D. Daniels, O. Farkas, J. B. Foresman, J. V. Ortiz, J. Cioslowski, and D. J. Fox, Gaussian, Inc., Wallingford CT, 2010.

[32] Gaussian 03, Revision C.02, M. J. Frisch, G. W. Trucks, H. B. Schlegel, G. E. Scuseria, M. A. Robb, J. R. Cheeseman, J. A. Montgomery, Jr., T. Vreven, K. N. Kudin, J. C. Burant, J. M. Millam, S. S. Iyengar, J. Tomasi, V. Barone, B. Mennucci, M. Cossi, G. Scalmani, N. Rega, G. A. Petersson, H. Nakatsuji, M. Hada, M. Ehara, K. Toyota, R. Fukuda, J. Hasegawa, M. Ishida, T. Nakajima, Y. Honda, O. Kitao, H. Nakai, M. Klene, X. Li, J. E. Knox, H. P. Hratchian, J. B. Cross, C. Adamo, J. Jaramillo, R. Gomperts, R. E. Stratmann, O. Yazyev, A. J. Austin, R. Cammi, C. Pomelli, J. W. Ochterski, P. Y. Ayala, K. Morokuma, G. A. Voth, P. Salvador, J. J. Dannenberg, V. G. Zakrzewski, S. Dapprich, A. D. Daniels, M. C. Strain, O. Farkas, D. K. Malick, A. D. Rabuck, K. Raghavachari, J. B. Foresman, J. V. Ortiz, Q. Cui, A. G. Baboul, S. Clifford, J. Cioslowski, B. B. Stefanov, G. Liu, A. Liashenko, P. Piskorz, I. Komaromi, R. L. Martin, D. J. Fox, T. Keith, M. A. Al-Laham, C. Y. Peng, A. Nanayakkara, M. Challacombe, P. M. W. Gill, B. Johnson, W. Chen, M. W. Wong, C. Gonzalez, and J. A. Pople, Gaussian, Inc., Wallingford CT, 2004.

[33] Parr, R.G. and Weitao, Y. *Density Functional Theory of Atoms and Molecules*; Oxford University Press: New York, 1989; pp. 87-104.

[34] Nelson Flores-Gallegos and Carmen Salazar-Hernández. *Some Applications of Quantum Mechanics; Chapter 10, Flows of Information and Informational Trajectories in Chemical Processes*. InTech, 2012; pp. 233-256.

Quantum Effects Through a Fractal Theory of Motion

M. Agop, C.Gh. Buzea, S. Bacaita, A. Stroe and M. Popa

Additional information is available at the end of the chapter

1. Introduction

Scale Relativity Theory (SRT) affirms that the laws of physics apply in all reference systems, whatever its state of motion and its scale. In consequence, SRT imply [1-3] the followings:

i.   Particle movement on continuous and non-differentiable curve (or almost nowhere differentiable), that is explicitly scale dependent and its length tends to infinity, when the scale interval tends to zero.

ii.   Physical quantities will be expressed through fractal functions, namely through functions that are dependent both on coordinate field and resolution scale. The invariance of the physical quantities in relation with the resolution scale generates special types of transformations, called resolution scale transformations. In what follows we will explain the above statement.

Let $F(x)$ be a fractal function in the interval $x \in [a, b]$ and let the sequence of values for x be:

$$x_a = x_0, x_1 = x_0 + \varepsilon, x_k = x_0 + k\varepsilon, x_n = x_0 + n\varepsilon = x_b \tag{1}$$

We can now say that $F(x, \varepsilon)$ is a –scale approximation.

Let us now consider as a $\bar{\varepsilon}$-scale approximation of the same function. Since $F(x)$ is everywhere almost self-similar, if ε and $\bar{\varepsilon}$ are sufficiently small, both approximations $F(x, \varepsilon)$ and must lead to same results. By comparing the two cases, one notices that scale expansion is related to the increase $d\varepsilon$ of ε, according to an increase $d\bar{\varepsilon}$ of $\bar{\varepsilon}$. But, in this case we have:

$$\frac{d\varepsilon}{\varepsilon} = \frac{d\bar{\varepsilon}}{\varepsilon} = d\rho \tag{2}$$

situation in which we can consider the infinitesimal scale transformation as being

$$\varepsilon' = \varepsilon + d\varepsilon = \varepsilon + \varepsilon d\rho \tag{3}$$

Such transformation in the case of function $F(x, \varepsilon)$, leads to:

$$F(x,\varepsilon') = F(x,\varepsilon + \varepsilon d\rho) \tag{4}$$

respectively, if we limit ourselves to a first order approximation:

$$F(x,\varepsilon') = F(x,\varepsilon) + \frac{\partial F(x,\varepsilon)}{\partial \varepsilon}(\varepsilon' - \varepsilon) = F(x,\varepsilon) + \frac{\partial F(x,\varepsilon)}{\partial \varepsilon}\varepsilon d\rho \tag{5}$$

Moreover, let us notice that for an arbitrary but fixed ε_0, we obtain:

$$\frac{\partial \ln(\varepsilon/\varepsilon_0)}{\partial \varepsilon} = \frac{\partial(\ln \varepsilon - \ln \varepsilon_0)}{\partial \varepsilon} = \frac{1}{\varepsilon} \tag{6}$$

situation in which (5) can be written as:

$$F(x,\varepsilon') = F(x,\varepsilon) + \frac{\partial F(x,\varepsilon)}{\partial \ln(\varepsilon/\varepsilon_0)}d\rho = \left[1 + \frac{\partial}{\partial \ln(\varepsilon/\varepsilon_0)}d\rho\right]F(x,\varepsilon) \tag{7}$$

Therefore, we can introduce the dilatation operator:

$$\hat{D} = \frac{\partial}{\partial \ln(\varepsilon/\varepsilon_0)} \tag{8}$$

At the same time, relation (8) shows that the intrinsic variable of resolution is not ε, but $\ln(\varepsilon / \varepsilon_0)$.

The fractal function is explicitly dependent on the resolution $(\varepsilon / \varepsilon_0)$, therefore we have to solve the differential equation:

$$\frac{dF}{d\ln(\varepsilon/\varepsilon_0)} = P(F) \tag{9}$$

where $P(F)$ is now an unknown function. The simplest explicit suggested form for $P(F)$ is linear dependence [2]

$$P(F) = A + BF, A, B = const. \tag{10}$$

in which case the differential equation (9) takes the form:

$$\frac{dF}{d\ln(\varepsilon/\varepsilon_0)} = A + BF \tag{11}$$

Hence by integration and substituting:

$$B = -\tau, \tag{12}$$

$$-\frac{A}{B} = F_0 \tag{13}$$

we obtain:

$$F\left(\frac{\varepsilon}{\varepsilon_0}\right) = F_0\left[1 + \left(\frac{\varepsilon_0}{\varepsilon}\right)^{\tau}\right] \tag{14}$$

We can now generalize the previous result by considering that F is dependent on parameterization of the fractal curve. If p characterizes the position on the fractal curve then, following the same algorithm as above, the solution will be as a sum of two terms i.e. both classical and differentiable (depending only on position) and fractal, non-differentiable (depending on position and, divergently, on $\varepsilon/\varepsilon_0$)

$$F(p, \varepsilon/\varepsilon_0) = F_0(p)\left[1 + \xi(p)\left(\frac{\varepsilon_0}{\varepsilon}\right)^{\tau(p)}\right] \tag{15}$$

where $\xi(p)$ is a function depending on parameterization of the fractal curve.

The following particular cases are to be considered:

1. in asymptotic small scale regime $\varepsilon \langle\langle \varepsilon_0,$ τ is constant (with no scale dependence) and power-law dependence on resolution is obtained:

$$F(p, \varepsilon/\varepsilon_0) = T(p)\left(\frac{\varepsilon_0}{\varepsilon}\right)^{\tau} \qquad \text{a}$$

$$T(p) = F_0(p)Q(p) \qquad \text{b}$$

$$\text{(16)}$$

2. in the asymptotic big scale regime $\varepsilon \rangle\rangle\varepsilon_0$, τ is constant (with no scale dependence) and, in terms of resolution, one obtains an independent law:

$$F(p, \varepsilon/\varepsilon_0) \rightarrow F_0(p) \qquad \text{(17)}$$

Particularly, if $F(p, \varepsilon/\varepsilon_0)$ are the coordinates in given space, we can write

$$X(p, \varepsilon/\varepsilon_0) = x(p)\left[1 + \xi(p)\left(\frac{\varepsilon_0}{\varepsilon}\right)^{\tau}\right] \qquad \text{(18)}$$

In this situation, $\xi(p)$ becomes a highly fluctuating function which can be described by stochastic process while τ represents (according to previous description) the difference between fractal and topological dimensions. The result is a sum of two terms, a classical, differentiable one (dependent only on the position) and a fractal, non-differentiable one (dependent both on the position and, divergently, on $\varepsilon/\varepsilon_0$). This represents the importance of the above analysis.

By differentiating these two parts we obtain:

$$dX = dx + d\xi \qquad \text{(19)}$$

where dx is the classical differential element and $d\xi$ is a differential fractal one.

iii. There is infinity of fractal curves (geodesics) relating to any couple of points (or starting from any point) and applied for any scale. The phenomenon can be easily understood at the level of fractal surfaces, which, in their turn, can be described in terms of fractal distribution of conic points of positive and negative infinite curvature. As a consequence, we have replaced velocity on a particular geodesic by fractal velocity field of the whole infinite ensemble of geodesics. This representation is similar to that of fluid mechanics [4] where the motion of the fluid is described in terms of its velocity field $v = (x(t), t)$, density $\rho = (x(t), t)$ and, possibly, its pressure. We shall, indeed, recover the fundamental equations of fluid mechanics (Euler and continuity equations), but we shall write them in terms of a density of probability (as defined by the set of geodesics) instead of a density of matter and adding an additional term of quantum pressure (the expression of fractal geometry).

iv.   The local differential time invariance is broken, so the time-derivative of the fractal field Q can be written two-fold:

$$
\frac{d_+Q}{dt} = \lim_{\Delta t \to 0_+} \frac{Q(t+\Delta t)-Q(t)}{\Delta t} \qquad \text{a}
$$
$$
\frac{d_-Q}{dt} = \lim_{\Delta t \to 0_-} \frac{Q(t)-Q(t-\Delta t)}{\Delta t} \qquad \text{b}
$$
(20)

Both definitions are equivalent in the differentiable case $dt \to -dt$. In the non-differentiable situation, these definitions are no longer valid, since limits are not defined anymore. Fractal theory defines physics in relationship with the function behavior during the "zoom" operation on the time resolution δt, here identified with the differential element dt (substitution principle), which is considered an independent variable. The standard field $Q(t)$ is therefore replaced by fractal field $Q(t,dt)$, explicitly dependent on time resolution interval, whose derivative is not defined at the unnoticeable limit $dt \to 0$. As a consequence, this leads to the two derivatives of the fractal field Q as explicit functions of the two variables t and dt,

$$
\frac{d_+Q}{dt} = \lim_{\Delta t \to 0_+} \frac{Q(t+\Delta t,\Delta t)-Q(t,\Delta t)}{\Delta t} \qquad \text{a}
$$
$$
\frac{d_-Q}{dt} = \lim_{\Delta t \to 0_-} \frac{Q(t,\Delta t)-Q(t-\Delta t,\Delta t)}{\Delta t} \qquad \text{b}
$$
(21)

Notation "+" corresponds to the forward process, while "-" to the backward one.

v.   We denote the average of these vectors by $dx_\pm^i$, i.e.

$$
\left\langle dX_\pm^i \right\rangle = dx_\pm^i, i = 1,2
$$
(22)

Since, according to (19), we can write:

$$
dX_\pm^i = dx_\pm^i + d\xi_\pm^i
$$
(23)

and it results:

$$
\left\langle d\xi_\pm^i \right\rangle = 0
$$
(24)

vi.   The differential fractal part satisfies the fractal equation:

$$d_{\pm}\xi^i = \lambda_{\pm}^i (dt)^{1/D_F} \tag{25}$$

where $\lambda_{\pm}^i$ are some constant coefficients and D_F is a constant fractal dimension. We note that the use of any Kolmogorov or Hausdorff [1, 5, 6-8] definitions can be accepted for fractal dimension, but once a certain definition is admitted, it should be used until the end of analyzed dynamics.

vii.   The local differential time reflection invariance is recovered by combining the two derivatives, d_+/dt and d_-/dt, in the complex operator:

$$\frac{\hat{d}}{dt} = \frac{1}{2}\left(\frac{d_+ + d_-}{dt}\right) - \frac{i}{2}\left(\frac{d_+ - d_-}{dt}\right) \tag{26}$$

Applying this operator to the "position vector", a complex velocity yields

$$\hat{V} = \frac{\hat{d}X}{dt} = \frac{1}{2}\left(\frac{d_+X + d_-X}{dt}\right) - \frac{i}{2}\left(\frac{d_+X - d_-X}{dt}\right) = \frac{V_+ + V_-}{2} - i\frac{V_+ - V_-}{2} = V - iU \tag{27}$$

with:

$$V = \frac{V_+ + V_-}{2} \qquad \text{a}$$
$$\tag{28}$$
$$U = \frac{V_+ - V_-}{2} \qquad \text{b}$$

The real part, V, of the complex velocity $\hat{V}$, represents the standard classical velocity, which does not depend on resolution, while the imaginary part, U, is a new quantity coming from resolution dependant fractal.

2. Covariant total derivative

Let us now assume that curves describing particle movement (continuous but non-differentiable) are immersed in a 3-dimensional space, and that X of components $X^i (i = \overline{1, 3})$ is the position vector of a point on the curve. Let us also consider a fractal field $Q(X, t)$ and expand its total differential up to the third order:

$$d_{\pm}Q = \frac{\partial Q}{\partial t}dt + \nabla Q \cdot d_{\pm}X + +\frac{1}{2}\frac{\partial^2 Q}{\partial X^i \partial X^j}d_{\pm}X^i d_{\pm}X^j + \frac{1}{6}\frac{\partial^3 Q}{\partial X^i \partial X^j \partial X^k}d_{\pm}X^i d_{\pm}X^j d_{\pm}X^k \tag{29}$$

where only the first three terms were used in Nottale's theory (*i.e.* second order terms in the motion equation). Relations (29) are valid in any point both for the spatial manifold and for the points X on the fractal curve (selected in relations 29). Hence, the forward and backward average values of these relations take the form:

$$\langle d_{\pm}Q \rangle = \left\langle \frac{\partial Q}{\partial t} dt \right\rangle + \langle \nabla Q \cdot d_{\pm}X \rangle + \frac{1}{2}\left\langle \frac{\partial^2 Q}{\partial X^i \partial X^j} d_{\pm}X^i d_{\pm}X^j \right\rangle + \frac{1}{6}\left\langle \frac{\partial^3 Q}{\partial X^i \partial X^j \partial X^k} d_{\pm}X^i d_{\pm}X^j d_{\pm}X^k \right\rangle \tag{30}$$

The following aspects should be mentioned: the mean value of function f and its derivatives coincide with themselves and the differentials $d_{\pm}X^i$ and dt are independent; therefore, the average of their products coincides with the product of averages. Consequently, the equations (30) become:

$$d_{\pm}Q = \frac{\partial Q}{\partial t} dt + \nabla Q \langle d_{\pm}X \rangle + \frac{1}{2}\frac{\partial^2 Q}{\partial X^i \partial X^j} \langle d_{\pm}X^i d_{\pm}X^j \rangle + \frac{1}{6}\frac{\partial^3 Q}{\partial X^i \partial X^j \partial X^k} \langle d_{\pm}X^i d_{\pm}X^j d_{\pm}X^k \rangle \tag{31}$$

or more, using equations (23) with characteristics (24),

$$d_{\pm}Q = \frac{\partial Q}{\partial t} dt + \nabla Q \cdot d_{\pm}X + \frac{1}{2}\frac{\partial^2 Q}{\partial X^i \partial X^j}\left(d_{\pm}x^i d_{\pm}x^j + \langle d_{\pm}\xi^i d_{\pm}\xi^j \rangle \right) +$$
$$\frac{1}{6}\frac{\partial^3 Q}{\partial X^i \partial X^j \partial X^k}\left(d_{\pm}x^i d_{\pm}x^j d_{\pm}x^k + \langle d_{\pm}\xi^i d_{\pm}\xi^j d_{\pm}\xi^k \rangle \right) \tag{32}$$

Even if the average value of the fractal coordinate $d_{\pm}\xi^i$ is null (see 24), for higher order of fractal coordinate average, the situation can still be different. Firstly, let us focus on the averages $\langle d_{+}\xi^i d_{+}\xi^j \rangle$ and $\langle d_{-}\xi^i d_{-}\xi^j \rangle$. If $i \neq j$, these averages are zero due to the independence of $d_{\pm}\xi^i$ and $d_{\pm}\xi^j$. So, using (25), we can write:

$$\langle d_{\pm}\xi^i d_{\pm}\xi^j \rangle = \lambda_{\pm}^i \lambda_{\pm}^j \left(dt \right)^{(2/D_F)-1} dt \tag{33}$$

Then, let us consider the averages $\langle d_{\pm}\xi^i d_{\pm}\xi^j d_{\pm}\xi^k \rangle$. If $i \neq j \neq k$, these averages are zero due to independence of $d_{\pm}\xi^i$ on $d_{\pm}\xi^j$ and $d_{\pm}\xi^k$. Now, using equations (25), we can write:

$$\langle d_{\pm}\xi^i d_{\pm}\xi^j d_{\pm}\xi^k \rangle = \lambda_{\pm}^i \lambda_{+}^j \lambda_{\pm}^k \left(dt \right)^{(3/D_F)-1} dt \tag{34}$$

Then, equations (32) may be written as follows:

$$d_\pm Q = \frac{\partial Q}{\partial t} dt + d_\pm \mathbf{x} \cdot \nabla Q + \frac{1}{2} \frac{\partial^2 Q}{\partial X^i \partial X^j} d_\pm x^i d_\pm x^j + \frac{1}{2} \frac{\partial^2 Q}{\partial X^i \partial X^j} \lambda_\pm^i \lambda_\pm^j (dt)^{(2/D_F)-1} dt +$$

$$\frac{1}{6} \frac{\partial^3 Q}{\partial X^i \partial X^j \partial X^k} d_\pm x^i d_\pm x^j d_\pm x^k + \frac{1}{6} \frac{\partial^3 Q}{\partial X^i \partial X^j \partial X^k} \lambda_\pm^i \lambda_\pm^j \lambda_\pm^k (dt)^{(3/D_F)-1} dt \tag{35}$$

If we divide by dt and neglect the terms containing differential factors (for details on the method see [9, 10]), equations (38a) and (38b) are reduced to:

$$\frac{d_\pm Q}{dt} = \frac{\partial Q}{\partial t} + V_\pm \cdot \nabla Q + \frac{1}{2} \frac{\partial^2 Q}{\partial X^i \partial X^j} \lambda_\pm^i \lambda_\pm^j (dt)^{(2/D_F)-1} + \frac{1}{6} \frac{\partial^3 Q}{\partial X^i \partial X^j \partial X^k} \lambda_\pm^i \lambda_\pm^j \lambda_\pm^k (dt)^{(3/D_F)-1} \tag{36}$$

These relations also allow us to define the operator:

$$\frac{d_\pm}{dt} = \frac{\partial}{\partial t} + V_\pm \cdot \nabla + \frac{1}{2} \frac{\partial^2}{\partial X^i \partial X^j} \lambda_\pm^i \lambda_\pm^j (dt)^{(2/D_F)-1} + \frac{1}{6} \frac{\partial^3}{\partial X^i \partial X^j \partial X^k} \lambda_\pm^i \lambda_\pm^j \lambda_\pm^k (dt)^{(3/D_F)-1} \tag{37}$$

Under these circumstances, let us calculate $\left(\hat{\partial} Q / \partial t \right)$. Taking into account equations (26), (27) and (37), we shall obtain:

$$\frac{\hat{\partial} Q}{\partial t} = \frac{1}{2} \left[\frac{d_+ Q}{dt} + \frac{d_- Q}{dt} - i \left(\frac{d_+ Q}{dt} - \frac{d_- Q}{dt} \right) \right] =$$

$$= \frac{1}{2} \frac{\partial Q}{\partial t} + \frac{1}{2} V_+ \cdot \nabla Q + \lambda_+^i \lambda_+^j \frac{1}{4} (dt)^{(2/D_F)-1} \frac{\partial^2 Q}{\partial X^i \partial X^j} + \lambda_+^i \lambda_+^j \lambda_+^k \frac{1}{12} (dt)^{(3/D_F)-1} \frac{\partial^3 Q}{\partial X^i \partial X^j \partial X^k} +$$

$$+ \frac{1}{2} \frac{\partial Q}{\partial t} + \frac{1}{2} V_- \cdot \nabla Q + \lambda_-^i \lambda_-^j \frac{1}{4} (dt)^{(2/D_F)-1} \frac{\partial^2 Q}{\partial X^i \partial X^j} + \lambda_-^i \lambda_-^j \lambda_-^k \frac{1}{12} (dt)^{(3/D_F)-1} \frac{\partial^3 Q}{\partial X^i \partial X^j \partial X^k} -$$

$$- \frac{i}{2} \frac{\partial Q}{\partial t} - \frac{i}{2} V_+ \cdot \nabla Q - \lambda_+^i \lambda_+^j \frac{i}{2} (dt)^{(2/D_F)-1} \frac{\partial^2 Q}{\partial X^i \partial X^j} - \lambda_+^i \lambda_+^j \lambda_+^k \frac{i}{12} (dt)^{(3/D_F)-1} \frac{\partial^3 Q}{\partial X^i \partial X^j \partial X^k} +$$

$$+ \frac{i}{2} \frac{\partial Q}{\partial t} + \frac{i}{2} V_- \cdot \nabla Q + \lambda_-^i \lambda_-^j \frac{i}{2} (dt)^{(2/D_F)-1} \frac{\partial^2 Q}{\partial X^i \partial X^j} + \lambda_-^i \lambda_-^j \lambda_-^k \frac{i}{12} (dt)^{(3/D_F)-1} \frac{\partial^3 Q}{\partial X^i \partial X^j \partial X^k} = \tag{38}$$

$$= \frac{\partial Q}{\partial t} + \left(\frac{V_+ + V_-}{2} - i \frac{V_+ - V_-}{2} \right) \cdot \nabla Q + \frac{(dt)^{(2/D_F)-1}}{4} \left[\left(\lambda_+^i \lambda_+^j + \lambda_-^i \lambda_-^j \right) - i \left(\lambda_+^i \lambda_+^j - \lambda_-^i \lambda_-^j \right) \right] \frac{\partial^2 Q}{\partial X^i \partial X^j} +$$

$$+ \frac{(dt)^{(3/D_F)-1}}{12} \left[\left(\lambda_+^i \lambda_+^j \lambda_+^k + \lambda_-^i \lambda_-^j \lambda_-^k \right) - i \left(\lambda_+^i \lambda_+^j \lambda_+^k - \lambda_-^i \lambda_-^j \lambda_-^k \right) \right] \frac{\partial^3 Q}{\partial X^i \partial X^j \partial X^k} =$$

$$= \frac{\partial Q}{\partial t} + \hat{V} \cdot \nabla Q + \frac{(dt)^{(2/D_F)-1}}{4} \left[\left(\lambda_+^i \lambda_+^j + \lambda_-^i \lambda_-^j \right) - i \left(\lambda_+^i \lambda_+^j - \lambda_-^i \lambda_-^j \right) \right] \frac{\partial^2 Q}{\partial X^i \partial X^j} +$$

$$\frac{(dt)^{(3/D_F)-1}}{12} \left[\left(\lambda_+^i \lambda_+^j \lambda_+^k + \lambda_-^i \lambda_-^j \lambda_-^k \right) - i \left(\lambda_+^i \lambda_+^j \lambda_+^k - \lambda_-^i \lambda_-^j \lambda_-^k \right) \right] \frac{\partial^3 Q}{\partial X^i \partial X^j \partial X^k}$$

This relation also allows us to define the fractal operator:

$$\frac{\hat{\partial}}{\partial t} = \frac{\partial}{\partial t} + \hat{V} \cdot \nabla + \frac{(dt)^{(2/D_F)-1}}{4}\left[\left(\lambda_+^i\lambda_+^j + \lambda_-^i\lambda_-^j\right) - i\left(\lambda_+^i\lambda_+^j - \lambda_-^i\lambda_-^j\right)\right]\frac{\partial^2}{\partial X^i \partial X^j} +$$
$$+ \frac{(dt)^{(3/D_F)-1}}{12}\left[\left(\lambda_+^i\lambda_+^j\lambda_+^k + \lambda_-^i\lambda_-^j\lambda_-^k\right) - i\left(\lambda_+^i\lambda_+^j\lambda_+^k - \lambda_-^i\lambda_-^j\lambda_-^k\right)\right]\frac{\partial^3}{\partial X^i \partial X^j \partial X^k} \tag{39}$$

Particularly, by choosing:

$$\lambda_+^i\lambda_+^j = -\lambda_-^i\lambda_-^j = 2D\delta^{ij} \tag{40}$$

$$\lambda_+^i\lambda_+^j\lambda_+^k = -\lambda_-^i\lambda_-^j\lambda_-^k = 2\sqrt{2}D^{3/2}\delta^{ijk} \tag{41}$$

the fractal operator (39) takes the usual form:

$$\frac{\hat{\partial}}{\partial t} = \frac{\partial}{\partial t} + \hat{V} \cdot \nabla - i\mathbf{D}(dt)^{(2/D_F)-1}\Delta + \frac{\sqrt{2}}{3}\mathbf{D}^{3/2}(dt)^{(3/D_F)-1}\nabla^3 \tag{42}$$

We now apply the principle of scale covariance and postulate that the passage from classical (differentiable) to "fractal" mechanics can be implemented by replacing the standard time derivative operator, d/dt, with the complex operator $\partial/\partial t$ (this results in a generalization of Nottale's [1, 2] principle of scale covariance). Consequently, we are now able to write the diffusion equation in its covariant form:

$$\frac{\hat{\partial}Q}{\partial t} = \frac{\partial Q}{\partial t} + \left(\hat{V} \cdot \nabla\right)Q - i\mathbf{D}(dt)^{(2/D_F)-1}\Delta Q + \frac{\sqrt{2}}{3}\mathbf{D}^{3/2}(dt)^{(3/D_F)-1}\nabla^3 Q = 0 \tag{43}$$

This means that at any point on a fractal path, the local temporal $\partial_t Q$, the non-linear (convective), $\left(\hat{V} \cdot \nabla\right)Q$, the dissipative, ΔQ, and the dispersive, $\nabla^3 Q$, terms keep their balance.

3. Fractal space-time and the motion equation of free particles in the dissipative approximation

Newton's fundamental equation of dynamics in the dissipative approximation is:

$$m\frac{\partial \hat{V}}{\partial t} = -\nabla\Phi \tag{44}$$

where m is the mass, $\hat{V}$ the instantaneous velocity of the particle, Φ the scalar potential and

$$\frac{\hat{\partial}}{\partial t} = \frac{\partial}{\partial t} + \left(\hat{V}\cdot\nabla\right) - iD(dt)^{(2/D_F)-1}\Delta \tag{45}$$

is the fractal operator in the dissipative approximation.

In what follows, we study what happens with equation (44), in the free particle case ($\Phi = 0$), if one considers the space-time where particles move changes from classical to nondifferentiable.

According to Nottale [11], the transition from classical (differentiable) mechanics to the scale relativistic framework is implemented by passing to a fluid-like description (the fractality of space), considering the velocity field a fractal function explicitly depending on a scale variable (the fractal geometry of each geodesic). Separating the real and imaginary parts, (44) becomes:

$$\frac{\partial\mathbf{V}}{\partial t} + \mathbf{V}\cdot\nabla\mathbf{V} = 0$$
$$\mathbf{U}\cdot\nabla\mathbf{V} = -D\Delta\mathbf{V} \tag{46}$$

where $\hat{V} = V - i\,U$ is the complex velocity defined through (27) and D defines the amplitude of the fractal fluctuations ($D = D(dt)^{(2/D_F)-1}$).

Let us analyze in what follows, the second equation (46) which, one can see, may contain some interesting physics. If we compare it with Navier-Stokes equation, from fluid mechanics [12]

$$\frac{D\mathbf{v}}{Dt} = \frac{\partial\mathbf{v}}{\partial t} + \mathbf{v}\cdot\nabla\mathbf{v} = \nu\nabla^2\mathbf{v} \tag{47}$$

we can see the left side of (46) gives the rate at which V is transported through a 'fluid' by means of the motion of 'fluid' particles with the velocity U; the right hand side gives the diffusion of V, (D which is the amplitude of the fractal fluctuations, plays here the role of the 'cinematic viscosity' of the 'fluid'). One can notice, in those regions in which the right hand side of (47) is negligible, $Dv/Dt = 0$. This means that in inviscid flows, for instance, $\hat{V}$ is frozen into the 'particles of the fluid'. Physically this is due to the fact that in an inviscid 'fluid' shear stresses are zero, so that there is no mechanism by which $\hat{V}$ can be transferred from one 'fluid' particle to another. This may be the case for the transport of V by U in the second equation (II.3).

If we consider the flow of V induced by a uniform translational motion of a plane spaced a distance Y above a stationary parallel plane (Fig. 1), and if the 'fluid' velocity increases from zero (at the stationary plane) to U (at the moving plane) like in the case of simple Couette flow, or simple shear flow, then

$$\text{rate of shear deformation} = \frac{dV}{dy} = \frac{U}{Y} \tag{48}$$

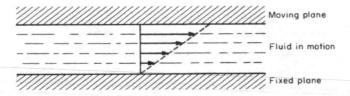

Figure 1. Uniform translational motion of a plane spaced a distance Y above a stationary parallel plane.

For many fluids it is found that the magnitude of the shearing stress is related to the rate of shear proportionally:

$$T = \eta \frac{dV}{dy} = \eta \frac{U}{Y} \tag{49}$$

Fluids which obey (49) in the above situation are known as *Newtonian fluids*, which have a very small coefficient of viscosity. When such 'fluids' flow at reasonable velocities it is found that viscous effects appear only in thin layers on the surface of objects or surfaces over which the 'fluid' flows. That is, if one continues the analogy, and questions how is V transported by the motion of 'fluid' particles with the velocity U, in second equation (46), one can assume that the mechanism of transfer of V from one particle of 'fluid' to another is achieved over small distances (in thin layers, as stated above).

We study an important case, of the one-dimensional flow along the Ox axis :

$$V = \zeta(x)\, k \tag{50}$$

To resume, the model considered here consists in analyzing the transport of V, along a small elementary distance Λ, by the 'particles' of a *Newtonian fluid* moving with velocity U, where the stress tensor obeys (49), i.e.

$$\frac{d\varsigma(x)}{dx} = \frac{\varsigma(x)}{\Lambda} \qquad (51)$$

like in the case of simple Couette flow, or simple shear flow.

Consequently, the second eq. (46) reduces to the scalar equation

$$\varsigma''(x) + K^2(x)\varsigma(x) = 0 \qquad (52)$$

which is the time independent Schrödinger equation, and

$$K^2(x) = \frac{1}{\Lambda D}U(x) \qquad (53)$$

with Λ and D having the significance of a small elementary distance and of the 'cinematic viscosity' (or amplitude of the fractal fluctuations), respectively, and $U(x)$ is the velocity of the 'Newtonian fluid', which is nothing but the imaginary part of the complex velocity [13]. In what follows, we solve this equation accurately by means of the WKBJ approximation method with connection formulas.

3.1. Solving the Schrödinger type equation by means of the WKBJ approximation method

Let us re-write (53) in the form

$$K^2(x) = \frac{1}{\Lambda D}U(x) = \frac{2m^2c}{\hbar^2}(\chi - \gamma(x)) \qquad (54)$$

where we take $D = \hbar/2m$ and consider the small elementary distance the Compton length $\Lambda = \hbar/mc$ [14]. Therefore, the Schrödinger equation (52) splits into:

$$\frac{d^2}{dx^2}\varsigma(x) + k^2(x)\varsigma(x) = 0, \ \chi > \gamma \ \text{or} \ \frac{d^2}{dx^2}\varsigma(x) - \rho^2(x)\varsigma(x) = 0, \ \chi < \gamma \qquad (55)$$

where

$$k(x) = \sqrt{\frac{2\mu(\chi - \gamma(x))}{\hbar^2}}, \rho(x) = \sqrt{\frac{2\mu(\gamma(x) - \chi)}{\hbar^2}} \ \text{with} \ \mu = m^2c \qquad (56)$$

χ is a limit velocity and $\gamma(x)$ a 'velocity potential'.

Let us try a solution of the form $\zeta(x)=A\ exp\ ((i/\hbar)S(x))$. Substituting this solution into the time-independent Schrödinger equation (52) we get:

$$i\hbar\frac{d^2S}{dx^2}-\left(\frac{dS}{dx}\right)^2+\hbar^2k^2=0 \text{ or } i\hbar\frac{d^2S}{dx^2}-\left(\frac{dS}{dx}\right)^2-\hbar^2\rho^2=0 \tag{57}$$

Assume that $\hbar$ can, in some sense, be regarded as a small quantity and that $S(x)$ can be expanded in powers of $\hbar$, $S(x) = S_0(x)+\hbar\ S_1(x) +....$

Then,

$$i\hbar\frac{d}{dx}\left[\frac{dS_0}{dx}+\hbar\frac{dS_1}{dx}+...\right]-\left(\frac{dS_0}{dx}+\hbar\frac{dS_1}{dx}+...\right)^2+\hbar^2k^2=0, \quad (\chi>\gamma(x)) \tag{58}$$

We assume that $\left|\dfrac{dS_0}{dx}\right|>>\left|\hbar\dfrac{dS_1}{dx}\right|$ and collect terms with equal powers of $\hbar$.

$$-\left[\frac{dS_0}{dx}\right]^2+\hbar^2k^2=0 \quad\Rightarrow\quad S_0=\pm\int_x\hbar k(x')dx' \tag{59}$$

$$i\frac{d^2S_0}{dx^2}-2\frac{dS_0}{dx}\frac{dS_1}{dx}=0 \quad\Rightarrow\quad S_1=\frac{1}{2}i\ln k(x) \tag{60}$$

We have used:

$$i\frac{d}{dx}\left(\frac{dS_0}{dx}\right)=2\frac{dS_0}{dx}\frac{dS_1}{dx}, \quad i\frac{dk}{dx}=2k\frac{dS_1}{dx}, \quad dS_1=\frac{i}{2}\frac{dk}{k} \tag{61}$$

Therefore, for $\chi>\gamma(x)$

$$\zeta(x)=Ak^{-\frac{1}{2}}e^{\pm i\int_x k(x')dx'} \tag{62}$$

In the classically allowed region $S_0=\pm\int_x\hbar k(x')dx'$ counts the oscillations of the velocity wave function. An increase of $2\pi\hbar$ corresponds to an additional phase of 2π.

Similarly, in regions where $\chi < \gamma(x)$ we have:

$$\zeta(x) = A\rho^{-\frac{1}{2}} e^{\pm\int_x \rho(x')dx'} \tag{63}$$

For our first order expansion to be accurate we need that the magnitude of higher order terms decreases rapidly. We need $\left| \frac{dS_0}{dx} \right| >> \left| \hbar \frac{dS_1}{dx} \right|$ or $|k| >> \left| \frac{1}{2k} \frac{dk}{dx} \right|$. The local deBroglie wavelength is $\lambda = 2\pi/k$. Therefore, $\left| \frac{\lambda}{4\pi} \frac{d\lambda}{dx} \right| << \lambda$, i.e. the change in λ over a distance $\lambda/4\pi$ is small compared to λ. This holds when the velocity potential $\gamma(x)$ varies slowly and the momentum is nearly constant over several wavelengths.

Near the classical turning points the WKBJ solutions become invalid, because k goes to zero here. We have to find a way to connect an oscillating solution to an exponential solution across a turning point if we want to solve barrier penetration problems or find bound states.

3.2. Velocity potential γ (x) and the bound states

We want to find the velocity wave function in a given velocity potential well $\gamma(x)$. Assuming that the limit velocity of the particle is χ and that the classical turning points are x_1 and x_2, $x_1 < x_2$, i.e. we have a velocity potential well with two sloping sides (Fig. 2).

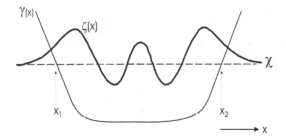

Figure 2. Bound state problem.

For $x < x_1$ the velocity wave function is of the form:

$$\zeta_1(x) = A_1 \rho^{-\frac{1}{2}} e^{\int_x \rho(x')dx'} \tag{64}$$

For $x > x_2$ the velocity wave function is of the form:

$$\zeta_3(x) = A_3 \rho^{-\frac{1}{2}} e^{-\int_x \rho(x')dx'} \tag{65}$$

In the region between x_1 and x_2 it is of the form:

$$\zeta_2(x) = A_2 k^{-\frac{1}{2}} e^{+i\int_x^{} k(x')dx'} + A_2' k^{-\frac{1}{2}} e^{-i\int_x^{} k(x')dx'} \tag{66}$$

At $x = x_1$ and $x = x_2$ the velocity wave function ζ and its derivatives have to be continuous. Near x_1 and x_2 we expand the velocity potential well $\gamma(x)$ in a Taylor series expansion in x and neglect all terms of order higher than 1. Near x_1 we have $\gamma(x) = \chi - K_1(x - x_1)$, and near x_2 we have $\gamma(x) = \chi + K_2(x - x_2)$.

In the neighborhood of x_1 the time-independent Schrödinger equation then becomes:

$$\frac{d^2\zeta}{dx^2} + \frac{2\mu K_1}{\hbar^2}(x - x_1)\zeta = 0 \tag{67}$$

and in the neighborhood of x_2 the time-independent Schrödinger equation becomes:

$$\frac{d^2\zeta}{dx^2} - \frac{2\mu K_2}{\hbar^2}(x - x_2)\zeta = 0 \tag{68}$$

Let us define $z = -\left(\frac{2\mu K_1}{\hbar^2}\right)^{\frac{1}{3}}(x - x_1)$. Then we obtain $\frac{d^2\zeta}{dz^2} - z\zeta = 0$ near x_1. The solutions of this equation which vanish asymptotically as $z \to \infty$ or $x \to -\infty$ are the Airy functions. They are defined through:

$$Ai(z) = \frac{1}{\pi}\int_0^\infty \cos\left(\frac{s^3}{3} + sz\right)ds \tag{69}$$

which for large $|z|$ has the asymptotic form

$$Ai(z) \sim \frac{1}{2\sqrt{\pi}z^{\frac{1}{4}}}\exp\left(-\frac{2}{3}z^{\frac{3}{2}}\right), \quad (z > 0) \tag{70}$$

and

$$Ai(z) \sim \frac{1}{\sqrt{\pi}(-z)^{\frac{1}{4}}}\sin\left(\frac{2}{3}(-z)^{\frac{3}{2}} + \frac{\pi}{4}\right), \quad (z < 0) \tag{71}$$

If the limit velocity χ is high enough, the linear approximation to the velocity potential well remains valid over many wavelengths. The Airy functions can therefore be the connecting velocity wave functions through the turning point at x_1.

If we define $z = \left(\dfrac{2\mu K_2}{\hbar^2}\right)^{\frac{1}{3}}(x - x_2)$ then we find $\dfrac{d^2\zeta}{dz^2} - z\zeta = 0$ near $x = x_2$ and the Airy functions can also be the connecting velocity wave functions through the turning point at x_2. Here $z \to \infty$ or $x \to \infty$.

In the neighborhood of x_1 we have

$$k^2 = -\rho^2 = \left(\frac{2\mu K_1}{\hbar^2}\right)^{\frac{1}{3}}(x - x_1) = -\left(\frac{2\mu K_1}{\hbar^2}\right)^{\frac{1}{3}} z \tag{72}$$

Therefore

$$\int_{x_1}^{x} \rho dx' = \left(\frac{2\mu K_1}{\hbar^2}\right)^{\frac{1}{3}} \int_{x_1}^{x} \sqrt{z}\,dx' = -\int_{0}^{x}\sqrt{z'}\,dz' = -\frac{2}{3}z^{\frac{3}{2}} \tag{73}$$

Similarly

$$\int_{x_1}^{x} k dx' = \left(\frac{2\mu K_1}{\hbar^2}\right)^{\frac{1}{3}} \int_{x_1}^{x} \sqrt{-z}\,dx' = -\int_{0}^{x}\sqrt{-z'}\,dz' = \frac{2}{3}(-z)^{\frac{3}{2}} \tag{74}$$

By comparing this with the asymptotic forms of the Airy functions we note that

$$\zeta_1(x) = A_1\rho^{-\frac{1}{2}}e^{+\int_{x_1}^{x}\rho(x')dx'} \qquad (x < x_1) \tag{75}$$

must continue on the right side as

$$\zeta_2(x) = 2A_1 k^{-\frac{1}{2}} \sin\left(\int_{x_1}^{x} k dx' + \frac{\pi}{4}\right)(x > x_1) \tag{76}$$

In the neighborhood of x_2 we similarly find that

$$\zeta_3(x) = A_3 \rho^{-\frac{1}{2}} e^{-\int_{x_2}^{x} \rho(x')dx'} \quad (x > x_2) \tag{77}$$

must continue in region 2 as

$$\zeta_2(x) = 2A_3 k^{-\frac{1}{2}} \sin\left(\int_{x}^{x_2} kdx' + \frac{\pi}{4}\right) (x < x_2) \tag{78}$$

Both expressions for $\zeta_2(x)$ are approximations to the same eigenfunction. We therefore need

$$2A_1 k^{-\frac{1}{2}} \sin\left(\int_{x_1}^{x} kdx' + \frac{\pi}{4}\right) = 2A_3 k^{-\frac{1}{2}} \sin\left(\int_{x}^{x_2} kdx' + \frac{\pi}{4}\right) \tag{79}$$

For (79) to be satisfied, the amplitudes of each side must have the same magnitude, and the phases must be the same modulo π :

$$|A_1| = |A_3|$$
$$\int_{x_1}^{x} kdx' + \frac{\pi}{4} = -\int_{x}^{x_2} kdx' - \frac{\pi}{4} + n\pi \tag{80}$$

Knowing that $\int_{x_1}^{x_2} = \int_{x_1}^{x} + \int_{x}^{x_2}$, we have

$$\int_{x_1}^{x_2} kdx' = \left(n - \frac{1}{2}\right)\pi, \quad n = 1,2,3,... \tag{81}$$

This can be re-written as

$$\int_{x_1}^{x_2} \Pi dx = \left(n - \frac{1}{2}\right)\frac{h}{2} \text{ or } \oint \Pi dx = \left(n - \frac{1}{2}\right)h \tag{82}$$

with

$$\Pi = \left[2\mu\big(\chi - \gamma(x)\big) \right]^{1/2} = \left[2\mu U(x) \right]^{1/2} = mc \left[2\frac{U(x)}{c} \right]^{1/2} \tag{83}$$

Here $\oint$ denote an integral over one complete cycle of the classical motion. The WKBJ method for $\gamma(x)$ velocity potential well with soft walls, therefore, leads to a Wilson-Sommerfeld type quantization rule except that n is replaced by $n-1/2$. It leads to a quantization of the complex velocity $U(x)$.

The factor of $\pi/2$ arises here due to the two phase changes of $\pi/4$ at x_1 and x_2. In case where only one of the walls is soft and the other is infinitely steep the factor of $1/2$ is replaced by $1/4$ in (81). If both walls are infinitely steep, the factor of $1/2$ in (81) is replaced by 0.

WKBJ approximation is a semi classical approximation, since it is expected to be most useful in the nearly classical limit of large quantum numbers. The method will not be good for, say, lowest limit velocity states χ, so in order to overcome this shortcomings there is a need for a modified semi classical quantization condition. For oscillations between the two classical turning points x_1 and x_2, we obtain the semi classical quantization condition by requiring that the total phase during one period of oscillation to be an integral multiple of 2π; [15] such that

$$2 \int_{x_1}^{x_2} k\,dx' + \phi_1 + \phi_2 = 2\pi n \tag{84}$$

where ϕ_1 is the phase loss due to reflection at the classical turning point x_1 and ϕ_2 is the phase loss due to reflection at x_2. Taking ϕ_1 and ϕ_2 to be equal to $\pi/2$ leads to the modified semiclassical quantization rule, i.e.

$$\int_{x_1}^{x_2} k\,dx' = \left(n - \frac{m}{4} \right)\pi \tag{85}$$

where m is the Maslov index [15], which denotes the total phase loss during one period in units of $\pi/2$. It contains contributions from the phase losses ϕ_1 and ϕ_2 due to reflections at points x_1 and x_2, respectively. It is pertinent to note that taking $\phi_1 = \phi_2 = \pi/2$ and an integer Maslov index $m = 2$ in (85), we have the familiar semi classical quantization rule, i.e. (81).

Let us apply the constraint equation (81) to an harmonic oscillator. The condition then is (passing without loss of generality to the limits $-a$ to $+a$)

$$\int_{-a}^{+a}\left[2\mu\left(\chi-\omega x\right)\right]^{1/2} dx = \left(n-\frac{1}{2}\right)\pi\hbar \tag{86}$$

where the energy W_c of the oscillator $U(x)$ with the pulsation ω writes

$$W_c = \frac{1}{2}m\omega^2 x^2 = \frac{1}{2}m\gamma^2(x) \tag{87}$$

and we get the expression for the x dependence of the velocity term, $\gamma(x)=\omega x$.

The left side term of (86) is an elementary integral and we find:

$$\left(\chi+\upsilon\right)^{3/2} - \left(\chi-\upsilon\right)^{3/2} = \frac{3\pi}{2a\left(2\mu\right)^{1/2}}\left(n-\frac{1}{2}\right)\hbar\upsilon \tag{88}$$

where $\upsilon=\omega\,a$ is the liniar velocity (see the graphic in Fig. 3).

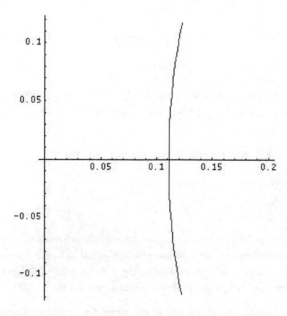

Figure 3. Dependence of the limit velocity χ on the linear velocity υ.

We try to estimate a value for the limit velocity χ. Let us expand the left side term of (88) in series and keep the first term. If we replace μ from (56) and take $a = \Lambda$ (the Compton length), we get:

$$\chi \approx \frac{\pi^2 \hbar^2}{8\mu a^2}\left(n - \frac{1}{2}\right)^2 = \frac{\pi^2}{32}(2n-1)^2 c = \left\{\frac{\pi^2}{32}c, \ 9\frac{\pi^2}{32}c, \ 25\frac{\pi^2}{32}c, \ \ldots\right\} \tag{89}$$

It is interesting to note that only the first velocity in (89) is less than the velocity of light, c.

Let us analyze now, one more bound state, the velocity wave function in a given velocity double well potential $\gamma(x)$.

We begin by deriving a quantization condition for region 2 analogous to (81). Again, applying the boundary condition for region 1 leaves only the exponentially growing solution. Applying the connection formula at x_1 then gives an expression for the velocity wave function in region 2:

$$\varsigma_1(x) = 2Ak^{-1/2} \sin\left[\int_{x_1}^{x} kdx' + \frac{\pi}{4}\right] \tag{90}$$

However, the solution in region 3 must have both growing and decaying solutions present. Considering the region 3 solutions in terms of x_2 and letting B_L and C_L be the amplitudes of the decaying and growing solutions respectively, the connection formulas give another expression for the velocity wave function in region 2:

$$\zeta_2(x) = 2B_L k^{-1/2}\cos\theta + C_L k^{-1/2}\sin\theta \tag{91}$$

with

$$\theta = \int_{x}^{x_2} kdx' - \frac{\pi}{4} \tag{92}$$

We equate the two expressions (90), (91) for the velocity function in region 2 and cancel common factors giving

$$2A\sin\left[\int_{x_1}^{x} kdx' + \frac{\pi}{4}\right] = 2B_L \cos\theta + C_L \sin\theta \tag{93}$$

Using trigonometric identities to simplify the right hand side, gives

$$2A \sin\left[\int_{x_1}^{x} k dx' + \frac{\pi}{4}\right] = \left(4B_L^2 + C_L^2\right)^{1/2} \sin\left(\theta + \frac{\pi}{2} - \phi_L\right) \tag{94}$$

where

$$\phi_L = \cos^{-1}\left[\frac{2B_L}{\left(4B_L^2 + C_L^2\right)^{1/2}}\right] \tag{95}$$

The magnitude of the sin function must be equal, and the magnitude of the phases must be equal modulo π :

$$4A^2 = 4B_L^2 + C_L^2 \tag{96}$$

$$\int_{x_1}^{x} k dx' + \frac{\pi}{4} = -\int_{x}^{x_2} k dx' - \frac{\pi}{4} + \phi_L + n\pi \tag{97}$$

Simplifying and combining the integrals gives the quantization condition for region 2:

$$\theta_{12} \equiv \int_{x_1}^{x_2} k dx = \pi\left(n - \frac{1}{2}\right) + \phi_L \tag{98}$$

with $n = 1, 2,....$

A similar treatment for the turning point x_3 yields the condition for region 4:

$$\theta_{34} \equiv \int_{x_3}^{x_4} k dx = \pi\left(m - \frac{1}{2}\right) + \phi_R \tag{99}$$

with $m = 1, 2,...$ and ϕ_R given by:

$$\phi_R = \cos^{-1}\left[\frac{2C_R}{\left(4C_R^2 + B_R^2\right)^{1/2}}\right] \tag{100}$$

where B_R and C_R are the amplitudes of the decaying and growing region 3 solutions in terms of x_3.

We now have the quantization conditions (98, 99) for regions 2 and 4, but they contain the free parameters ϕ_L and ϕ_R. To eliminate these free parameters, we consider the WKBJ solution in region 3. The coefficients B_L, C_L, B_R, C_R define two expressions for solution, which must be equal:

$$\zeta_3 = B_L \rho^{-1/2} \exp\left[-\int_{x_2}^{x} \rho dx'\right] + C_L \rho^{-1/2} \exp\left[\int_{x_2}^{x} \rho dx'\right] \tag{101}$$

$$\zeta_3 = B_R \rho^{-1/2} \exp\left[\int_{x}^{x_3} \rho dx'\right] + C_R \rho^{-1/2} \exp\left[-\int_{x}^{x_3} \rho dx'\right] \tag{102}$$

Equations (101) and (102) each contain a term that grows exponentially with x and a term that decays exponentially with x. Equating the growing terms from each equation and the decaying term from each equation gives two constraints:

$$B_L \exp\left[-\int_{x_2}^{x} \rho dx'\right] = B_R \exp\left[\int_{x}^{x_3} \rho dx'\right] \tag{103}$$

$$C_L \exp\left[\int_{x_2}^{x} \rho dx'\right] = C_R \exp\left[-\int_{x}^{x_3} \rho dx'\right] \tag{104}$$

Combining the integrals in these constraints gives

$$\frac{B_L}{B_R} = \frac{C_R}{C_L} = \exp(\theta_{23}) \tag{105}$$

with

$$\theta_{23} \equiv \int_{x_2}^{x_3} \rho dx' \tag{106}$$

The constraints (98, 99, 105) may be combined to give a single quantization condition for the allowed WKBJ velocity limits χ for a double-well velocity potential $\gamma(x)$. Applying trigonometric identities to (95) and (100), and plugging into (105) gives

$$\tan\phi_L \tan\phi_R = \left(\frac{C_L}{2B_L}\right)\left(\frac{B_R}{2C_R}\right) = \frac{1}{4}\exp\left(-2\theta_{23}\right) \tag{107}$$

Equation (107) may be combined with (98) and (99) to give the WKBJ quantization condition for a double-well potential in terms of the phase integrals θ_{12} and θ_{34} :

$$ctg\theta_{12}ctg\theta_{34} = \frac{1}{4}\exp\left(-2\theta_{23}\right) \tag{108}$$

confirming the results given in [16].

Equation (108) is a nonlinear constraint approximately determining the allowed velocity levels χ of a double-well velocity potential $\gamma(x)$ (see Fig. 4) and can be written (taking $\phi_R = \phi_L = \pi/4$ in (98) and (99), i.e. the velocity quarter-wave shift in the connection formulas, which is known to optimize the tunneling effect between two oscillating waves [17]) as :

$$\frac{1}{\hbar}\int_{x_2}^{x_3}\left[2\mu\left(\gamma(x)-\chi\right)\right]^{1/2}dx = \ln\left\{4\cdot ctg\left[\pi\left(n-\frac{1}{2}\right)+\phi_L\right]\cdot ctg\left[\pi\left(m-\frac{1}{2}\right)+\phi_R\right]\right\}^{-1/2} = -\ln 2 \quad m,n=1,2,3,\dots \tag{109}$$

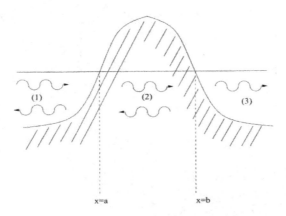

Figure 4. Tunneling potential barrier.

In terms of the momentum Π we have :

$$\int_{x_2}^{x_3} \Pi dx = -\hbar \ln 2 \text{ or } \oint \Pi dx = -2\hbar \ln 2 \qquad (110)$$

where $\oint$ denotes an integral over one complete cycle of the classical motion, this time

$$\Pi = \left[2\mu\left(\gamma\left(x\right) - \chi\right)\right]^{1/2} = \left[2\mu U\left(x\right)\right]^{1/2} = mc\left[2\frac{U\left(x\right)}{c}\right]^{1/2} \qquad (111)$$

since $\gamma(x) > \chi$ for the integration limits, i.e. region 3 (see Fig. 5). We get again a quantization of the complex velocity $U(x)$, where the levels are equally spaced at a value of $\hbar \ln 2$.

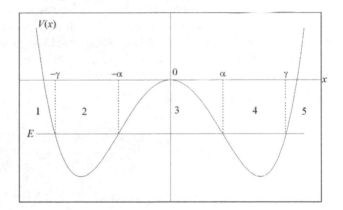

Figure 5. Schematic diagram of a double-well potential with three forbidden regions (1, 3, 5) and two allowed regions (2, 4).

In 1961, Landauer [18] discussed the limitation of the efficiency of a computer imposed by physical laws. In particular he argued that, according to the second law of thermodynamics, the erasure of one bit of information requires a minimal heat generation $k_B T \ln 2$, where k_B is Boltzmann's constant and T is the temperature at which one erases. Its argument runs as follows. Since erasure is a logical function that does not have a single-valued inverse it must be associated with physical irreversibility and therefore requires heat dissipation. A bit has one degree of freedom and so the heat dissipation should be of order $k_B T$. Now, since before erasure a bit can be in any of the two possible states and after erasure it can only be in one state, this implies a change in information entropy of an amount $-k_B \ln 2$.

The one-to-one dynamics of Hamiltonian systems [19] implies that when a bit is erased the information which it contains has to go somewhere. If the information goes into observable degrees of freedom of the computer, such as another bit, then it has not been erased but merely moved; but if it goes into unobservable degrees of freedom such as the microscopic motion of molecules it results in an increase of entropy of at least $k_B\ln2$.

Inspired by such studies, a considerable amount of work has been made on the thermodynamics of information processing, which include Maxwell's demon problem [20], reversible computation [21], the proposal of the algorithmic entropy [22] and so on.

Here, considering a double-well velocity potential $\gamma(x)$ and the velocity quarter-wave shift in the connection formulas, a quanta of $\hbar\ln2$ for the complex velocity $U(x)$ of the moving Newtonian 'fluid' occurs. It can be argued that it can be put into a one-to-one correspondence to the quanta of information Landauer and other authors discussed about [23, 24].

Furthermore, one gets an interesting result when taking $\phi_R = \phi_L = \pi/2$, i.e. the velocity half-wave shift in the connection formulas, when singularities occur in (II.66). We try to solve this case by making use of the vortices theory. Benard in 1908 was the first to investigate the appearance of vortices behind a body moving in a fluid [12]. The body he used was a cylinder. He observed that at a high enough fluid velocity (or Reynolds number based on the cylinder diameter), which depends on the viscosity and width of the body, vortices start to shed behind the cylinder, alternatively from the top and the bottom of the cylinder.

Consequently, we write (109) in the form

$$\int_x \left[2\mu\big(\gamma(x)-\chi\big)\right]^{1/2} dx = -\hbar\ln2 - \hbar\ln\left\{tg\left[\frac{\pi}{l}(x-x_0)\right]\right\} \tag{112}$$

where we use $ctg(\alpha + \pi/2) = - tg(\alpha)$, take m = n, make the notations $x = nl$, $x_0 = l/2$ and consider again the one-dimensional case, motion along the Ox axis.

Solving (112) one gets

$$U(x) = c\left(1 + ctg^2\left[\frac{2\pi}{\Lambda}(x-x_0)\right]\right) \tag{113}$$

where we assume $l = \Lambda$ (the Compton length), $U(x) = [\gamma(x) - \chi]/2\pi^2$ and replace $\mu = m^2c$, where c is the velocity of light. When plotting (113) (see Fig. 6) we see that indeed, singularities are obtained for $x - x_0 = \Lambda/2$ and for $x - x_0 = \Lambda/4$ we get for $U(x)$ minima of value the velocity of light, c.

Usually, at some distance behind a body placed in a fluid, vortices are arranged at a definite distance l apart and with a definite separation h between the two rows. The senses of the rotation in the two rows are opposite (see Fig. 7).

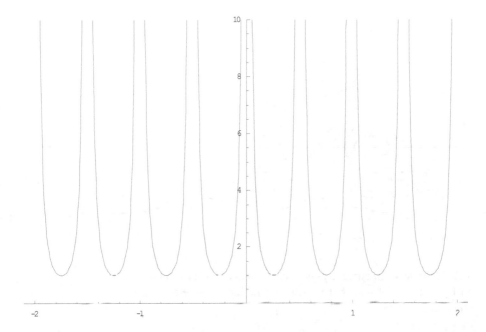

Figure 6. The complex velocity U(X) singularities' distribution along the Ox axis.

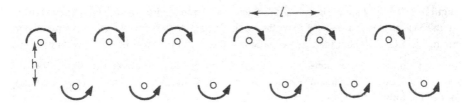

Figure 7. Von Karman vortex streets.

In 1912 von Karman expounded a theory of such vortex streets and the drag which a cylinder would experience due to their formation [12]. Since we considered here the one-dimensional case, we get the solution of a single row of rectilinear vortices, which has already been referred to as characterizing a surface of discontinuity (see Fig. 8).

A typical bound state in a double-well velocity potential has two classically allowed regions, where the velocity potential $\gamma(x)$ is less than the limit velocity χ. These regions are separated by a classically forbidden region, or barrier, where the velocity potential is larger than the limit velocity. As we can see, quantum mechanics predicts that a velocity wave $\zeta(x)$ travelling in such a potential is most likely to be found in the allowed regions. However, unlike classical mechanics, quantum mechanics predicts that this velocity wave can also be found in the

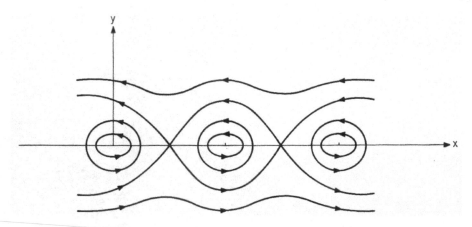

Figure 8. Single row of rectilinear vortices.

forbidden region. This uniquely quantum mechanical behavior allows a velocity wave, initially localized in one potential well, to penetrate through the barrier, into the other well (as we will see in what follows).

3.3. Velocity potential γ (x) and the quantum barrier

We already know at the points where $\chi-\gamma(x)=0$, special treatment is required because k is singular. The way of handling the solution near the turning point is a little bit more technical, but the basic idea is that we have a solution to the left and to the right of the turning point, and one needs a formula that interpolates between them. In other words, in the vicinity of the turning point one approximates $\sqrt{2\mu(\chi-\gamma(x))/\hbar^2}$ by a straight line over a small interval and solves TISE (time independent Schrodinger equation) exactly. This leads to the following connection formulas:

Barrier to the right (x = b turning point)

$$\frac{2}{\sqrt{k}}\cos\left[\int_x^b k(x)dx-\frac{\pi}{4}\right]\Leftrightarrow\frac{1}{\sqrt{\rho}}e^{-\int_b^x \rho(x)dx} \tag{114}$$

$$\frac{1}{\sqrt{k}}\sin\left[\int_x^b k(x)dx-\frac{\pi}{4}\right]\Leftrightarrow-\frac{1}{\sqrt{\rho}}e^{\int_b^x \rho(x)dx} \tag{115}$$

Barrier to the left (x = a turning point)

$$\frac{2}{\sqrt{k}}\cos\left[\int_a^x k(x)dx - \frac{\pi}{4}\right] \Leftrightarrow \frac{1}{\sqrt{\rho}}e^{-\int_x^a \rho(x)dx} \tag{116}$$

$$\frac{1}{\sqrt{k}}\sin\left[\int_a^x k(x)dx - \frac{\pi}{4}\right] \Leftrightarrow -\frac{1}{\sqrt{\rho}}e^{\int_x^a \rho(x)dx} \tag{117}$$

The connection formulas enable us to obtain relationships between the solutions in a region at some distance to the right of the turning point with those in a region at some distance to the left [25-27].

One of the most important problems to which connection formulas apply is that of the penetration of a potential barrier. The barrier is shown in Fig. 4 and the limit velocity χ is such that the turning points are at $x = a$ and $x = b$.

Suppose that the motion is incident from the left. Some waves will be reflected and some transmitted, so that in region III we will have:

$$\zeta_3(x) = \frac{1}{\sqrt{k}}e^{i\int_b^x kdx - i\frac{\pi}{4}} = \frac{1}{\sqrt{k}}\cos\left[\int_b^x kdx - \frac{\pi}{4}\right] + \frac{i}{\sqrt{k}}\sin\left[\int_b^x kdx - \frac{\pi}{4}\right] \tag{118}$$

The phase factor is included for convenience of applying the connection formulas.

In region II (using (114) and (115) on (118)) we have:

$$\zeta_2(x) = \frac{1}{2}\frac{1}{\sqrt{\rho}}e^{-\int_x^b \rho dx} - i\frac{1}{\sqrt{\rho}}e^{\int_x^b \rho dx} \tag{119}$$

Now using

$$\int_x^b \rho dx = \int_x^a \rho dx + \int_a^b \rho dx = -\int_a^x \rho dx + \alpha \tag{120}$$

we can write

$$\zeta_2(x) = \frac{1}{2}\frac{1}{\sqrt{\rho}}e^{\int_a^x \rho dx}e^{-\alpha} - i\frac{1}{\sqrt{\rho}}e^{-\int_a^x \rho dx}e^{\alpha} \tag{121}$$

Again, using the connection formulas for the case barrier to the right (using (116) and (117) on (121)), we get for region I:

$$\zeta_1(x) = \frac{1}{2}e^{-\alpha}\frac{1}{\sqrt{k}}\sin\left[\int_x^a kdx - \frac{\pi}{4}\right] - i\frac{2}{\sqrt{k}}e^{\alpha}\cos\left[\int_x^a kdx - \frac{\pi}{4}\right] =$$

$$= \frac{1}{2}\frac{1}{\sqrt{k}}e^{-\alpha}\sin(u) - i\frac{2}{\sqrt{k}}e^{\alpha}\cos(u) = \qquad (122)$$

$$= -\frac{i}{\sqrt{k}}\left[e^{iu}\left(e^{\alpha} + \frac{1}{4}e^{-\alpha}\right) + e^{-iu}\left(e^{\alpha} - \frac{1}{4}e^{-\alpha}\right)\right]$$

Hence

$$\begin{cases} \zeta_1^{inc}(x) = -\frac{i}{\sqrt{k}}\left(e^{\alpha} + \frac{1}{4}e^{-\alpha}\right)e^{i\int_x^a kdx - i\frac{\pi}{4}} \\ \\ \zeta_1^{ref}(x) = -\frac{i}{\sqrt{k}}\left(e^{\alpha} - \frac{1}{4}e^{-\alpha}\right)e^{-i\int_x^a kdx + i\frac{\pi}{4}} \end{cases} \qquad (123)$$

Having obtained the expression for $\zeta_1^{inc}(x)$ and $\zeta_1^{ref}(x)$ we are now in position to calculate the transmission coefficient using:

$$T = \left|\frac{\zeta_3(x)}{\zeta_1^{inc}(x)}\right|^2 = \frac{e^{-2\alpha}}{\left(1 + \frac{1}{4}e^{-2\alpha}\right)^2} \qquad (124)$$

To summarize, for a barrier with large attenuation $e^{-2\alpha} \to 0$, the tunneling probability equals

$$T = \frac{e^{-2\alpha}}{\left(1 + \frac{1}{4}e^{-2\alpha}\right)^2} \approx e^{-2\alpha} = \exp\left(-\frac{2}{\hbar}\int_a^b pdx\right) = \exp\left(-\frac{2}{\hbar}\int_a^b\left[2\mu\left(\gamma(x) - \chi\right)\right]^{1/2}dx\right) \qquad (125)$$

The reflection coefficient is:

$$R = \left|\frac{\zeta_1^{ref}(x)}{\zeta_1^{inc}(x)}\right|^2 \cong \frac{\left(e^{\alpha} - \frac{1}{4}e^{-\alpha}\right)^2}{\left(e^{\alpha} + \frac{1}{4}e^{-\alpha}\right)^2}, \quad T + R = 1 \qquad (126)$$

and also in the same large attenuation limit, we have:

$$R \approx 1 - e^{-2\alpha} = 1 - \exp\left(-\frac{2}{\hbar}\int_a^b \rho dx\right) = 1 - \exp\left(-\frac{2}{\hbar}\int_a^b \left[2\mu\left(\gamma(x) - \chi\right)\right]^{1/2} dx\right) \tag{127}$$

One can see from (125) and (127) thatthe velocity wave $\zeta(x)$ on small distances, with the same order of magnitude as Λ, may be influenced by $U(x)$,i.e. it can be transmitted, attenuated or reflected at this scale length. In other words, we get from the calculus, that the velocity field V is indeed transported by the motion of the 'Newtonian fluid' particles with the velocity $U(x)$ (the imaginary part of the complex velocity [13]).

4. Casimir type effect in scale relativity theory

In recent years, new and exciting advances in experimental techniques [28] prompted a great revival of interest in the Casimir effect, over fifty years after its theoretical discovery (for a recent review on both theoretical and experimental aspects of the Casimir effect, see Refs. [29-31]). As is well known, this phenomenon is a manifestation of the zero-point fluctuations of the electromagnetic field: it is a purely quantum effect and it constitutes one of the rare instances of quantum phenomena on a macroscopic scale.

In his famous paper, Casimir evaluated the force between two parallel, electrically neutral, perfectly reflecting plane mirrors, placed a distance L apart, and found it to be attractive and of a magnitude equal to:

$$F_C = \frac{\hbar c \pi^2 A}{240 L^4} \tag{128}$$

Here, A is the area of the mirrors, which is supposed to be much larger than L^2, so that edge effects become negligible. The associated energy E_C

$$E_C = -\frac{\hbar c \pi^2 A}{720 L^3} \tag{129}$$

can be interpreted as representing the shift in the zero-point energy of the electromagnetic field, between the mirrors, when they are adiabatically moved towards each other starting from an infinite distance. The Casimir force is indeed the dominant interaction between neutral bodies at the micrometer or submicrometer scales, and by modern experimental techniques it has now been measured with an accuracy of a few percent (see [28] and references therein).

Since this effect arises from long-range correlations between the dipole moments of the atoms forming the walls of the cavity, that are induced by coupling with the fluctuating electromagnetic field, the Casimir energy depends in general on the geometric features of the cavity. For example, we see from (129) that, in the simple case of two parallel slabs, the Casimir energy E_C is negative and is not proportional to the volume of the cavity, as would be the case for an extensive quantity, but actually depends separately on the area and distance of the slabs. Indeed, the dependence of E_C on the geometry of the cavity can reach the point where it turns from negative to positive, leading to repulsive forces on the walls. For example [29], in the case of a cavity with the shape of a parallelepiped, the sign of E_C depends on the ratios among the sides, while in the case of a sphere it has long been thought to be positive. It is difficult to give a simple intuitive explanation of these shape effects, as they hinge on a delicate process of renormalization, in which the finite final value of the Casimir energy is typically expressed as a difference among infinite positive quantities. In fact, there exists a debate, in the current literature, whether some of these results are true or false, being artifacts resulting from an oversimplification in the treatment of the walls [33].

There are three well-known technical types of derivation of the Casimir force for different geometries including the simplest geometry of two parallel, uncharged, perfectly conducting plates firstly explored by Casimir. One modern method is the quantum field theoretical approach based on the appropriate Green's function of the geometry of problem [34]. The other technical type is the dimensional regularization method that involves the mathematical complications of the Riemann zeta function and the analytical continuation [34]. The last (the most elementary/the simplest) method is based on modes summation by using the Euler-Maclurian integral formula [35-37].

The problem of finding the Casimir force, not only for the simplest geometry of two plates or rectangular prism, that we want to study here, but also for other more complicated geometries, indispensably/automatically involves some infinities/irregularities; thus, one should regularize the calculation for arriving at the desired finite physical result(s). In the Green' function method, one uses the subtraction of two terms (two Green's functions) to do the required regularization. In the dimensional regularization method, although there isn't an explicit subtraction for the regularization of the problem, as is clear from its name, the calculation is regularized dimensionally by going to a complex plane with a mathematically complicated/ambiguous approach. In the simplest method in which the Euler-Maclurian formula is used, the regularization is performed by the subtraction of the zero-point energy of the free space (no plates) from the energy expression under consideration/calculation (e.g. summation of the interior and exterior zero-point energies of the two parallel plates).

Navier-Stokes equations in scale relativity theory predict that the (vector) velocity field V and/or the (scalar) density field ρ, on small distances (the same magnitude as the Compton length) behave like a wave function and are transported by the motion of the Newtonian fluid with velocity U.

Furthermore, when considering vacuum from the Casimir cavity, a non-differentiable, Newtonian, 2D non-coherent quantum fluid whose entities (cvasi-particles) assimilated to vortex-type objects, initially non-coherent, become coherent (the coherence of the quantum fluid reduces to its ordering in vortex streets) due to the constraints induced by the presence of slabs. Casimir type forces are derived which are in good agreement with other theoretical results and experimental data, for both cases: two metallic slabs, parallel to each other, placed at a distance d apart, that constitute the plates of the cavity and a rectangle of sides d_1, d.

In other words, non-differentiability and coherence of the quantum fluid due to constraints generate pressure along the Ox and Oy axis.

For viscous compressible fluids, Navier-Stokes equations

$$\rho\frac{Dv}{Dt} = \rho X - \nabla p + \mu \nabla^2 v + \frac{\mu}{3}\nabla(\nabla \cdot v) \tag{130}$$

together with the equation of continuity

$$\frac{D\rho}{Dt} + \rho\nabla \cdot v = 0, \tag{131}$$

where ρ is the density, v the velocity of the fluid, X the body force, p the pressure, μ the shear viscosity and $D/Dt \equiv d/dt + v \cdot \nabla$ the Eulerian derivative, apply to *Newtonian* (or near) *fluids*, that is, to fluids in which the stress is linearly related to the rate of strain (as will be assumed further in this section) [12].

Let us see first, what happens with the set of equations (130) and (131), if one considers that the space-time, where particles move, changes from classical to non-differentiable.

We already know, according to Nottale [11], that a transition from classical (differentiable) mechanics to the scale relativistic framework is implemented by passing to a fluid-like description (the fractality of space), considering the velocity field a fractal function explicitly depending on a scale variable (the fractal geometry of each geodesic) and defining two fractal velocity fields which are fractal functions of the scale variable dt (the non-differentiability of space).

Consequently, replacing d/dt with the fractal operator (42) and solving for both real and imaginary parts, (130) and (131) become, in a stationary isotropic case, taking the body force $X = 0$ (constant gravitational field) and $\nabla U = 0$ (assuming a constant density of states for the "fluid particles" moving with the velocity U – see further in this section):

$$V \cdot \nabla V = -\frac{\nabla p}{\rho} + \upsilon \nabla^2 V \qquad \text{a} \tag{132}$$
$$U \cdot \nabla V + D\nabla^2 V = 0 \qquad \text{b}$$

and

$$V \cdot \nabla \rho + \rho \nabla \cdot V = 0 \qquad \text{a}$$
$$U \cdot \nabla \rho + D \nabla^2 \rho = 0 \qquad \text{b} \tag{133}$$

where V represents the standard classical velocity, which does not depend on resolution, while the imaginary part, U, is a new quantity coming from resolution dependant fractal, $v = \mu/\rho$ the kinematic viscosity and $D = \hbar/2m$ defines the amplitude of the fractal fluctuations.

The causes of the Casimir effect are described by quantum field theory, which states that all of the various fundamental fields, such as the electromagnetic field, must be quantized at each and every point in space. In a simplified view, a "field" in physics may be envisioned as if space were filled with interconnected vibrating balls and springs, and the strength of the field can be visualized as the displacement of a ball from its rest position. Vibrations in this field propagate and are governed by the appropriate wave equation for the particular field in question. The second quantization of quantum field theory requires that each such ball-spring combination to be quantized, that is, that the strength of the field to be quantized at each point in space. Canonically, the field at each point in space is a simple harmonic oscillator, and its quantization places a quantum harmonic oscillator at each point. Excitations of the field correspond to the elementary particles of particle physics. However, even the vacuum has a vastly complex structure, so all calculations of quantum field theory must be made in relation to this model of vacuum. The vacuum has, implicitly, all of the properties that a particle may have: spin, or polarization in the case of light, energy, and so on. On average, all of these properties cancel out: the vacuum is, after all, "empty" in this sense. One important exception is the vacuum energy or the vacuum expectation value of the energy.

Let us consider here, vacuum, as a non-differentiable, Newtonian, 2D non-coherent quantum fluid whose entities (cvasi-particles) assimilate to vortex-type objects [38] (see Fig.9) and are described by the wave function Ψ [39, 40]

$$\Psi = cn(\underline{u};k) \tag{134}$$

with

$$\underline{u} = \frac{K}{a}\underline{z}, \qquad \text{a}$$
$$\underline{z} = x + iy, \qquad \text{b}$$
$$\frac{K}{a} = \frac{K'}{b} \qquad \text{c}$$
$$K = \int_0^{\pi/2} \left(1 - k^2 \sin^2 \phi\right)^{-1/2} d\phi, \qquad \text{d} \tag{135}$$
$$K' = \int_0^{\pi/2} \left(1 - k'^2 \sin^2 \phi\right)^{-1/2} d\phi, \qquad \text{e}$$
$$k^2 + k'^2 = 1 \qquad \text{f}$$

and K, K' complete elliptic integrals of the first kind of modulus k [41], form a vortex lattice of constants a, b.

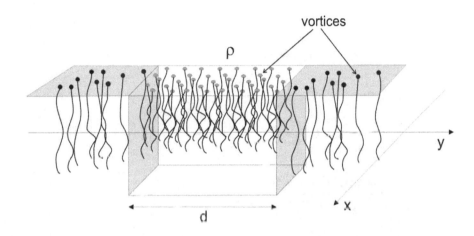

Figure 9. The vacuum from a Casimir cavity whose entities (cvasi-particles) are assimilated to vortex-type objects.

Applying in the complex plane [42], the formalism developed in [13] by means of the relation $\Psi = e^{F(z)/\Gamma} = cn(\underline{u};k)$ one introduces the complex potential

$$F(\underline{z}) = G(x,y) + iH(x,y) = \Gamma \ln\left[cn(\underline{u};k)\right] \tag{136}$$

with Γ the vortex constant. In the general case $\Gamma = c\Lambda = \hbar/m$ [38-40], the interaction scale being specified through Γ's value (Λ being considered as the Compton length).

Based on the complex potential (136), one defines the complex velocity field of the non-coherent quantum fluid, through the relation:

$$v_x - iv_y = \frac{dF(\underline{z})}{d\underline{z}} = -\frac{\Gamma K}{a}\frac{sn(\underline{u};k)dn(\underline{u};k)}{cn(\underline{u};k)} \tag{137}$$

or explicitly, using the notations [41, 42]:

$$
\begin{aligned}
s &= sn(\alpha,k), & \text{a} \\
c &= cn(\alpha,k), & \text{b} \\
d &= dn(\alpha,k), & \text{c} \\
\alpha &= \frac{K}{a}x, & \text{d} \\
s_1 &= sn(\beta,k'), & \text{e} \\
c_1 &= cn(\beta,k'), & \text{f} \\
d_1 &= dn(\beta,k'), & \text{g} \\
\beta &= \frac{K}{a}y & \text{h}
\end{aligned}
\tag{138}
$$

$$
v_x - iv_y = -\Gamma\frac{K}{a}\left\{ \frac{scd\left[c_1^2\left(d_1^2 + k^2 c^2 s_1^2\right) - s_1^2 d_1^2\left(d^2 c_1^2 - k^2 s^2\right)\right]}{(1 - d^2 s_1^2)(c^2 c_1^2 + s^2 d^2 s_1^2 d_1^2)} \right.
$$

$$
\left. + i\,\frac{s_1 c_1 d_1\left[c^2\left(d^2 c_1^2 - k^2 s^2\right) + s^2 d^2\left(d_1^2 + k^2 c^2 s_1^2\right)\right]}{(1 - d^2 s_1^2)(c^2 c_1^2 + s^2 d^2 s_1^2 d_1^2)} \right\}
\tag{139}
$$

Having in view that $cn(\underline{u} + \Omega) = cn(\underline{u})$, where $\Omega = 2(2m+1)K + 2niK'$ and $m,n = \pm1, \pm2...$, for $k \to 0$ and $k' \to 1$ limits, respectively, the quantum fluid, initially non-coherent (the amplitudes and phases of quantum fluid entities are independent) becomes coherent (the amplitudes and phases of quantum fluid entities are correlated [43]). In this context, from Fig. 10a,b of the equipotential curves $G(x_r, y_r) = $ const., for $k^2 = 0,1$, it results that the coherence of the quantum fluid reduces to its ordering in vortex streets - see Fig. III.2a for vortex streets aligned with the Ox axis and Fig. 10b for vortex streets aligned with the Oy axis. This process of ordering is achieved by generation of quasi-particles. Indeed, in the usual quantum mechanics the imaginary term ($i\Theta$) from the energy, i.e. $E = E_0 + i\Theta$, induces elementary excitations named resonances (for details see the collision theory [44]). Similarly, by extending the collision theory to the fractal space-time [1, 45], will imply that the presence of the imaginary term $H(x_r, y_r)$ in the potential $F(z)$ will generate quasi-particles, as well.

Now, writing the Navier-Stokes equation (132a) and the equation of continuity (133a) in scale relativity theory for constant density (incompressible fluids) in two dimensions, one gets

$$
\frac{\partial p}{\partial x} = \rho D\left(\frac{\partial^2 v_x}{\partial x^2} + \frac{\partial^2 v_x}{\partial y^2}\right) - \rho\left(v_x\frac{\partial v_x}{\partial x} + v_y\frac{\partial v_x}{\partial y}\right) \qquad \text{a}
$$

$$
\frac{\partial p}{\partial y} = \rho D\left(\frac{\partial^2 v_y}{\partial x^2} + \frac{\partial^2 v_y}{\partial y^2}\right) - \rho\left(v_x\frac{\partial v_y}{\partial x} + v_y\frac{\partial v_y}{\partial y}\right) \qquad \text{b}
\tag{140}
$$

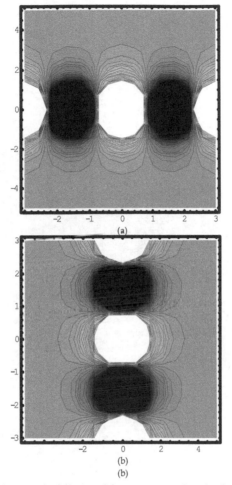

Figure 10. The equipotential curves $G(x_r,y_r)$ = const., a) for vortex streets aligned with the Ox axis and b) for vortex streets aligned with the Oy axis.

$$\frac{\partial v_x}{\partial x} + \frac{\partial v_y}{\partial y} = 0 \qquad (141)$$

where the shear viscosity v is replaced by D since we are dealing here with a non-differentiable quantum fluid.

Then, after some rather long yet elementary calculus one gets from (140a,b) through the degenerations :

i.

$$k=0, \quad k'=1, \quad K=\frac{\pi}{2}, \quad K'=\infty$$

$$p_y(\alpha) = -p_0 \sinh^2\left(\frac{\pi d}{2a}\right) \frac{1-\tan^2\alpha}{\cos(2\alpha)+\cosh\left(\frac{\pi d}{a}\right)} \qquad a$$

$$p_x(\beta) = -p_0 \sin^2\left(\frac{\pi d_1}{2a}\right) \frac{1+\tanh^2\beta}{\cos\left(\frac{\pi d_1}{a}\right)+\cosh(2\beta)} \qquad b$$

(142)

with

$$p_0 = \frac{\hbar^2 \pi^2 \rho}{4M^2 a^2}; \qquad a$$

$$\alpha = \frac{\pi x}{2a}; \qquad b$$

(143)

$$\beta = \frac{\pi y}{2a} \qquad c$$

and

ii.

$$k=1, \quad k'=0, \quad K=\infty, \quad K'=\frac{\pi}{2}$$

$$p_y(\alpha') = -p_0' \sin^2\left(\frac{\pi d}{2b}\right) \frac{1+\tanh^2\alpha'}{\cos\left(\frac{\pi d}{b}\right)+\cosh(2\alpha')} \qquad a$$

$$p_x(\beta') = -p_0' \sinh^2\left(\frac{\pi d_1}{2b}\right) \frac{1-\tan^2\beta'}{\cos(2\beta')+\cosh\left(\frac{\pi d_1}{b}\right)} \qquad b$$

(144)

with

$$p_0' = \frac{\hbar^2 \pi^2 \rho}{4M^2 b^2}; \qquad a$$

$$\alpha' = \frac{\pi x}{2b}; \qquad b_{\cdot}$$

(145)

$$\beta' = \frac{\pi y}{2b} \qquad c$$

Here, ρ is the quantum fluid's density, M the mass of the quantum fluid entities, d and d_1 are the elementary space intervals considered along the Oy and Ox axis, respectively.

In other words, non-differentiability and coherence of the quantum fluid due to constraints, generate pressure along the Ox and Oy axis.

Moreover, one can show that the equation of continuity (141) is identically satisfied for both cases of degeneration.

Let us consider a Casimir cavity consisting of the vacuum with the vortex lattice depicted above and two metallic slabs, that constitute the plates of the cavity, placed at a distance d apart, parallel to each other and to the xOz plane (see Fig. III.1). According to the analysis from the previous section, one can see that if the quantum fluid is placed in a potential well with infinite walls (the case of the Casimir cavity analyzed here, where the two plates are the constraints of the quantum fluid), along a direction perpendicular to the walls (the Oy axis here) a coherent structure, a vortex street forms (see Fig. III.2b). Consequently, by integrating (144a,b) with (145a-c) over α, and β_r, and using the result in the quantization rule:

$$\int_{d_1}^{d_2} kdx = n\pi, \qquad n = 1,2,3,... \tag{146}$$

where $d_1 \sim m\,\pi\,u, d \sim n\,\pi\,b$, with $m, n = 1,2,....$, one gets

$$\frac{\pi}{2}\frac{p_y}{p_0'} = 2r\arctan\left[\tan\left(\frac{n\pi^2}{2}\right)\tanh\left(\frac{m\pi^2}{4r}\right)\right]\tan^{-1}\left(n\pi^2\right) - r\tanh\left(\frac{m\pi^2}{4r}\right)$$

$$\frac{\pi}{2}\frac{p_x}{p_0'} = -2\arctan\left[\tan\left(\frac{n\pi^2}{4}\right)\tanh\left(\frac{m\pi^2}{2r}\right)\right]\tanh^{-1}\left(\frac{m\pi^2}{r}\right) + \tan\left(\frac{n\pi^2}{4}\right) \tag{147}$$

where

$$p_0' = \frac{\hbar^2\pi^2\rho}{4M^2b^2}; \qquad a$$

$$\alpha_r = \frac{\pi}{2}\frac{x}{a}; \qquad b$$

$$\beta_r = \frac{\pi}{2}\frac{y}{b}; \qquad c \tag{148}$$

$$r = \frac{b}{a} \qquad d$$

Graphically this is presented in Fig. III.3a,b for different values of the parameters $m, n = 1, 2,....$ and r.

If the plates were in the yOz plane the constraints being along the Ox axis, vortex streets would form along this axis and the result in (142a,b) with (143a-c) would have been applied, i.e. the cases i) or ii) are identical, yet they depend on the geometry chosen.

Firstly, one can notice that the pressure p_y on the plates, given by (147a), stabilizes for great r values, is always negative and an attractive force results (see Fig. 11 a), as is the case of the Casimir force (128).

Secondly, the theory predicts, that besides the pressure p_y acting on the plates, there must be yet another pressure, p_x (see Fig. 11 b), acting along the Ox axis and given by (147b). One can see that this pressure annuls for great r values, and has a minimum for some values of the parameters m, n. This result is new and should be checked by experiments.

Moreover, if one tries to compute the order of magnitude of this force, and replaces in (144a) : $\hbar = 1.054 \ 10^{-34}$ J.s, $m = 9.1 \ 10^{-31}$ kg, $\rho \sim 10^{21}$ cm^{-3}, $b = 1$Å (values specific to a bosonic gas, i.e. found in high-Tc superconductors [46]) and $d \sim 5 \, b$ (the distance between the plates), gets a value for $p_y \cong 6.18 \ 10^{10}$ N m^{-2} the same order of magnitude as the value calculated using (128), $F_c \cong 2.08 \ 10^{10}$ N m^{-2}.

As a final test, let us study the case of a Casimir cavity, as a rectangle of sides d_1, d. Now, the plates induce constraints along both Ox and Oy axis, thus correlations (vortex streets) form along these directions and one should use the degenerations i) and ii), simultaneously. Consequently, from (142a,b) with (143a-c) and (144a,b) with (145a-c) one gets

$$p_{y \, rect}\left(\alpha,\alpha'\right)=-\frac{\hbar^2\pi^2\rho}{4m^2}\left[\frac{1}{a^2}\frac{\sinh^2 A}{\cos^2\alpha}\left(1+\frac{\cosh 2A}{\cos 2\alpha}\right)^{-1}+\frac{1}{b^2}\frac{\sin^2 B}{\cosh^2\alpha'}\left(1+\frac{\cos 2B}{\cosh 2\alpha'}\right)^{-1}\right] \quad (149)$$

with

$$A=\frac{\pi d}{2a}; \quad B=\frac{\pi d}{2b} \quad (150)$$

and

$$p_{x \, rect}\left(\beta,\beta'\right)=-\frac{\hbar^2\pi^2\rho}{4m^2}\left[\frac{1}{a^2}\frac{\sin^2 A'}{\cosh^2\beta}\left(1+\frac{\cos 2A'}{\cosh 2\beta}\right)^{-1}+\frac{1}{b^2}\frac{\sinh^2 B'}{\cos^2\beta'}\left(1+\frac{\cosh 2B'}{\cos 2\beta'}\right)^{-1}\right] \quad (151)$$

with

$$A'=\frac{\pi d_1}{2a}; \quad B'=\frac{\pi d_1}{2b} \quad (152)$$

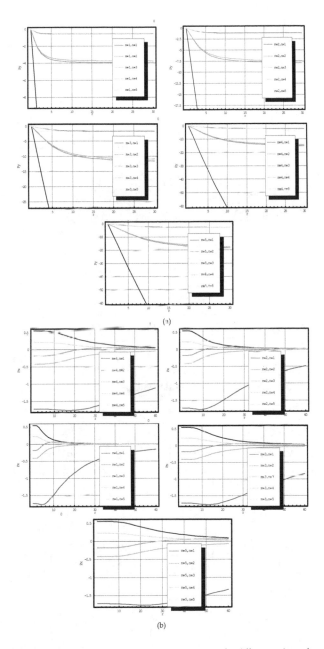

Figure 11. a) Plot of the pressure p_y on the plates, versus the parameter r for different values of parameters m, n; b) Plot of the pressure p_x versus the parameter r for different values of parameters m, n.

At every point (x, y) there is a pressure formed of the two constraints. Consequently, adding the pressures in (149) and (151) and using again the result in (146) (i.e. $d_1 \sim m \pi a, d \sim n \pi b$, where $m, n = 1,2,....$) one gets:

$$
-\frac{p_{rect}\left(\alpha_r, \beta_r\right)}{p_0'} = r^2 \frac{\sinh^2\left(\frac{n\pi^2}{2}r\right)}{\cos^2 \alpha_r}\left(1 + \frac{\cosh\left(n\pi^2 r\right)}{\cos\left(2\alpha_r\right)}\right)^{-1} + \frac{\sin^2\left(\frac{n\pi^2}{2}\right)}{\cosh^2\left(\frac{\alpha_r}{r}\right)}\left(1 + \frac{\cos\left(n\pi^2\right)}{\cosh\left(\frac{2\alpha_r}{r}\right)}\right)^{-1} +
$$

$$
+r^2 \frac{\sin^2\left(\frac{m\pi^2}{2}\right)}{\cosh^2\left(r\beta_r\right)}\left(1 + \frac{\cos\left(m\pi^2\right)}{\cosh\left(2r\beta_r\right)}\right)^{-1} + \frac{\sinh^2\left(\frac{1}{r}\frac{m\pi^2}{2}\right)}{\cos^2 \beta_r}\left(1 + \frac{\cosh\left(\frac{m\pi^2}{r}\right)}{\cos\left(2\beta_r\right)}\right)^{-1}
$$

$\qquad\qquad\qquad\qquad\qquad\qquad\qquad\qquad\qquad\qquad\qquad\qquad\qquad\qquad$ (153)

where

$$
p_0' = \frac{\hbar^2\pi^2\rho}{4M^2 b^2}; \quad \alpha_r = \frac{\pi}{2}\frac{x}{a} = \frac{\pi}{2}x_r; \quad \beta_r = \frac{\pi}{2}\frac{y}{b} = \frac{\pi}{2}y_r; \quad r = \frac{b}{a}
$$

$\qquad\qquad\qquad\qquad\qquad\qquad\qquad\qquad\qquad\qquad\qquad\qquad\qquad\qquad$ (154)

Furthermore, we integrate (153) over x_r and y_r, respectively, in order to find a value of the pressure acting on the sides of the rectangular enclosure. After some long, yet elementary calculus, one finds:

$$
\frac{p_{rect}}{p_0'} = -4nr^2 \frac{arctg\left[tg\left(\frac{m\pi^2}{4}\right)th\left(\frac{n\pi^2}{2}r\right)\right]}{th\left(n\pi^2 r\right)} + 4nr\frac{arctg\left[tg\left(\frac{n\pi^2}{2}\right)th\left(\frac{m\pi^2}{4}\frac{1}{r}\right)\right]}{tg\left(n\pi^2\right)} +
$$

$$
+2nr^2 tg\left(\frac{m\pi^2}{4}\right) - 2nr\, th\left(\frac{m\pi^2}{4}\frac{1}{r}\right) -
$$

$$
-4m\frac{arctg\left[tg\left(\frac{n\pi^2}{4}\right)th\left(\frac{m\pi^2}{2}\frac{1}{r}\right)\right]}{th\left(m\pi^2\frac{1}{r}\right)} + 4mr\frac{arctg\left[tg\left(\frac{m\pi^2}{2}\right)th\left(\frac{n\pi^2}{4}r\right)\right]}{tg\left(m\pi^2\right)} +
$$

$$
+2m\, tg\left(\frac{n\pi^2}{4}\right) - 2mr\, th\left(\frac{n\pi^2}{4}r\right)
$$

$\qquad\qquad\qquad\qquad\qquad\qquad\qquad\qquad\qquad\qquad\qquad\qquad\qquad\qquad$ (155)

Plots of (155) for various values of parameters $m, n = 1, 2,....$ and r are depicted in Fig. III.4a,b.

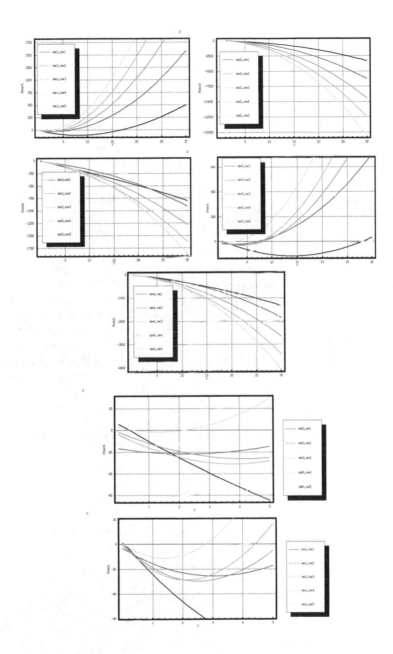

Figure 12. a) Plots of p_{rect} versus the parameter r for various values of parameters $m, n = 1,2,...$; b) the same plot, yet we present here a magnification of the domain of r for highly asymmetric values of m, n (1,5 and 5,1).

One can notice that if the two parameters m and n have close values, the force acting on the Casimir rectangle is always negative and decreases exponentially for increasing r. For parameters m and n (1,5 and 5,1, i.e. very asymmetric) the force has negative and positive domains (see Fig. 12 b) and increases exponentially for increasing r. Moreover, if one tries to find the positive and negative domains, and solve (155) for $m = 5$, $n = 5$ finds $p_{rect} < 0$ for $0.45753 \leq r \leq 2.18565$ and $p_{rect} > 0$ for $r > 2.18565$ and $r < 0.45753$. This result is in agreement with the calculus of regularization using the Abel-Plana formula where $E < 0$ for $0.36537 \leq L/l \leq 2.73686$ and $E > 0$ for $L/l > 2.73686$ and $L/l < 0.36537$ [47].

5. Fractal approximation of motion in mass transfer: release of drug from polimeric matrices

Polymer matrices can be produced in one of the following forms: micro/nano-particles, micro/nano capsules, hydro gels, films, patches.Our new approach considers the entire system (drug loaded polymer matrix in the release environment) as a type of "fluid" totally lacking interaction or neglecting physical interactions among particles. At the same time, the induced complexity is replaced by fractality. This will lead to particles moving on certain trajectories called geodesics within fractal space. This assumption represents the basis of the fractal approximation of motion in Scale Relativity Theory (SRT) [1, 2], leading to a generalized fractal "diffusion" equation that can be analyzed in terms of two approximations (dissipative and dispersive).

5.1. The dissipative approximation

In the dissipative approximation the fractal operator (42) takes the form [48, 49]:

$$\frac{\hat{\partial}}{\partial t} = \frac{\partial}{\partial t} + \hat{V} \cdot \nabla - iD(dt)^{(2/D_F)-1} \Delta \tag{156}$$

As a consequence, we are now able to write the fractal "diffusion" type equation in its covariant form:

$$\frac{\hat{\partial}Q}{dt} = \frac{\partial Q}{\partial t} + \left(\hat{V} \cdot \nabla\right)Q - iD(dt)^{(2/D_F)-1} \Delta Q = 0 \tag{157}$$

Separating the real and imaginary parts in (157), i.e.

$$\frac{\partial Q}{\partial t} + V \cdot \nabla Q = 0$$

$$-U \cdot \nabla Q = D(dt)^{(2/D_F)-1} \Delta Q \tag{158}$$

we can add these two equations and obtain a generalized "diffusion" type law in the form:

$$\frac{\partial Q}{\partial t} + (V - U) \cdot \nabla Q = D(dt)^{(2/D_F)-1} \Delta Q \tag{159}$$

5.1.1.Standard "diffusion" type equation. Fick type law

The standard "diffusion" law, *i.e.*:

$$\frac{\partial Q}{\partial t} = D \Delta Q \tag{160}$$

results from (159) on the following assertions:

i.   the diffusion path are the fractal curves of Peano's type. This means that the fractal dimension of the fractal curves is $D_F = 2$.

ii.   the movements at differentiable and non-differentiable scales are synchronous, i.e. $V = U$;

iii.   the structure coefficient D, proper to the fractal-nonfractal transition, is identified with the diffusion coefficient, i.e.

$D \equiv D$.

5.1.2. Anomalous "diffusion" type equation. Weibull relation

The anomalous diffusion law results from (IV.4) on the following assumptions:

i.   the diffusion path are fractal curves with fractal dimension $D_F \neq 2$;

ii.   the time resolution, δt, is identified with the differential element dt, i.e. the substitution principle can be applied also, in this case;

iii.   the movements at differentiable and non-differentiable scales are synchronous, i.e. $V = U$.

Then, the equation (IV.4) can be written:

$$\frac{\partial Q}{\partial t} = D(dt)^{(2/D_F)-1} \Delta Q \tag{161}$$

In one-dimensional case, applying the variable separation method [50]

$$Q(t,x) = T(t) \cdot X(x) \tag{162}$$

with the standard initial and boundary conditions:

$$Q(t,0) = 0, Q(t,L) = 0, Q(0,x) = F(x), 0 \le x \le L \tag{163}$$

implies:

$$\frac{1}{D(dt)^{(2/D_F)-1}} \frac{1}{T(t)} \frac{dT(t)}{dt} = \frac{1}{X(x)} \frac{d^2X(x)}{dx^2} = -m^2 = -\left(\frac{n\pi}{L}\right)^2, n = 1, 2 \tag{164}$$

where L is a system characteristic length, m a separation constant, dependent on diffusion order n.

Accepting the viability of the substitution principle, from (164), through integration, results:

$$\ln T = -m^2 D \int (dt)^{2/D_F} \tag{165}$$

Taking into consideration some results of the fractional integro-differential calculus [51, 52], (165) becomes:

$$\ln T = -\frac{m^2 D}{\Gamma\left(\frac{2}{D_F}+1\right)} t^{\frac{2}{D_F}}, \qquad a$$

$$\Gamma\left(\frac{2}{D_F}\right) = \int_0^\infty x^{\left(\frac{2}{D_F}\right)-1} e^{-x} dx \qquad b \tag{166}$$

Moreover, (166a,b) can be written under the form:

$$T(t) = \exp\left[-\frac{m^2 D}{\Gamma\left(\frac{2}{D_F}+1\right)} t^{\frac{2}{D_F}}\right] \tag{167}$$

The relative variation of concentrations, time dependent, is defined as:

$$T(t) = \frac{Q_\infty - Q_t}{Q_\infty} \tag{168}$$

where Q_t and Q_∞ are cumulative amounts of drug released at time t and infinite time.

From (167) and (168) results:

$$\frac{Q_t}{Q_\infty} = 1 - \exp\left[-\frac{m^2 D}{\Gamma\left(\frac{2}{D_F}+1\right)} t^{\frac{2}{D_F}} \right] \tag{169}$$

equation similar to Weibull relation $\frac{Q_t}{Q_\infty} = 1 - \exp(-at^b)$, a and b representing constants specific for each system that are defined by:

$$a - \frac{m^2 D}{\Gamma\left(\frac{2}{D_F}+1\right)} - \left(\frac{n\pi}{L}\right)^2 \frac{D}{\Gamma\left(\frac{2}{D_F}+1\right)} \qquad a$$

$$b = \frac{2}{D_F} \qquad\qquad b \tag{170}$$

We observe that both constants, a and b, are functions of the fractal dimension of the curves on which drug release mechanism take place, dimension that is a measure of the complexity and nonlinear dynamics of the system. Moreover, constant a depends, also, on the "diffusion" order n.

5.1.3. The correspondence between theoretical model and experimental results

The experimental and Weibull curves for HS (starch based hydrogels loaded with levofloxacin) and GA (GEL-PVA microparticles loaded with chloramphenicol) samples are plotted in Fig. 13.

The experimental data allowed to determine the values of Weibull parameters (a and b), and implicitly, the value of the fractal dimension from the curve on which release takes place [55].

These values confirmed that the complexity of the phenomena determines, also, naturally, a complex trajectory for the drug particles. Most values are between 1 and 3, in agreement with the values usually accepted for fractal process; higher values denotes the fact that, either fractal dimension must be redefined as function of structure "classes", or the drug release process is complex, involving many freedom degrees in the phase space [56]. Another observation that can be made based on this results is that the samples with $D_F \langle 2$ manifests a "sub-diffusion"

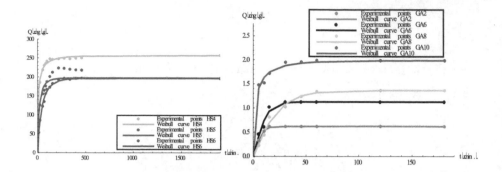

Figure 13. Experimental and Weibull curves for HS (left plot) and GA samples (right plot).

and, in the other, with $D_F \rangle 2$, the release process is of super-diffusion, classification in concordance with the experimental observation that this samples exhibit a "faster" diffusion, with a higher diffusion rate, in respect with the other samples [55].

5.2. The dispersive approximation

Let us now consider that, in comparison with dissipative processes, convective and dispersive processes are dominant ones. In these conditions, the fractal operator (42) takes the form:

$$\frac{\hat{\partial}}{dt} = \frac{\partial}{\partial t} + \left(\hat{V} \cdot \nabla\right) + \frac{\sqrt{2}}{3} D^{3/2} \left(dt\right)^{(3/D_{FD})-1} \nabla^3 \tag{171}$$

Consequently, we are now able to write the diffusion equation in its covariant form, as a Korteweg de Vries type equation:

$$\frac{\hat{\partial} Q}{dt} = \frac{\partial Q}{\partial t} + \left(\hat{V} \cdot \nabla\right) Q + \frac{\sqrt{2}}{3} D^{3/2} \left(dt\right)^{(3/D_{FD})-1} \nabla^3 Q = 0 \tag{172}$$

If we separate the real and imaginary parts from Eq. (172), we shall obtain:

$$\frac{\partial Q}{\partial t} + V \cdot \nabla Q + \frac{\sqrt{2}}{3} D^{3/2} \left(dt\right)^{(3/D_F)-1} \nabla^3 Q = 0 \qquad \text{a}$$
$$-U \cdot \nabla Q = 0 \qquad\qquad\qquad\qquad\qquad\qquad\qquad \text{b} \tag{173}$$

By adding them, the fractal diffusion equation is:

$$\frac{\partial Q}{\partial t} + (V - U) \cdot \nabla Q + \frac{\sqrt{2}}{3} D^{3/2} (dt)^{(3/D_F)-1} \nabla^3 Q = 0 \qquad (174)$$

From Eq. (173b) we see that, at fractal scale, there will be no Q field gradient.

Assuming that $|V - U| = \sigma \cdot Q$ with $\sigma =$ constant (in systems with self structuring processes, the speed fluctuations induced by fractal - non fractal are proportional with the concentration field [55]), in the particular one-dimensional case, equation (174) with normalized parameters:

$$
\begin{aligned}
\overline{\tau} &= \omega t, &\text{a} \\
\overline{\xi} &= kx, &\text{b} \\
\Phi &- \frac{Q}{Q_0} &\text{c}
\end{aligned}
\qquad (175)
$$

and normalizing conditions:

$$\frac{\sigma Q_0 k}{6\omega} = \frac{\sqrt{2}}{3} \frac{D^{3/2}(dt)^{(3/D_F)-1} k^3}{\omega} = 1 \qquad (176)$$

take the form:

$$\partial_{\overline{\tau}} \phi + 6\phi \partial_{\overline{\xi}} \phi + \partial_{\overline{\xi}\overline{\xi}\overline{\xi}} \phi = 0 \qquad (177)$$

In relations (175a,b,c) and (176) ω corresponds to a characteristic pulsation, k to the inverse of a characteristic length and Q_0 to balanced concentration.

Through substitutions:

$$
\begin{aligned}
w(\theta) &= \phi(\overline{\tau}, \overline{\xi}), &\text{a} \\
\theta &= \overline{\xi} - u\overline{\tau} &\text{b}
\end{aligned}
\qquad (178)
$$

eq.(177), by double integration, becomes:

$$\frac{1}{2} w'^2 = F(w) = -\left(w^3 - \frac{u}{2} w^2 - gw - h \right) \qquad (179)$$

with g, h two integration constants and u the normalized phase velocity. If $F(w)$ has real roots, equation (177) has the stationary solution:

$$\phi\left(\bar{\xi},\bar{\tau},s\right) = 2a\left(\frac{E(s)}{K(s)}-1\right) + 2a\cdot cn^2\left[\frac{\sqrt{a}}{s}\left(\bar{\xi}-\frac{u}{2}\bar{\tau}+\bar{\xi}_0\right);s\right] \tag{180}$$

where cn is Jacobi's elliptic function of s modulus [41], a is the amplitude, $\bar{\xi}_0$ is a constant of integration and

$$K(s) = \int_0^{\pi/2} \left(1-s^2\sin^2\phi\right)^{-1/2} d\phi \qquad a$$

$$\tag{181}$$

$$E(s) = \int_0^{\pi/2} \left(1-s^2\sin^2\phi\right)^{1/2} d\phi \qquad b$$

are the complete elliptic integrals [41].

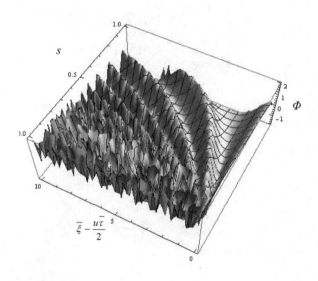

Figure 14. One-dimensional cnoidal oscillation modes of the field Φ

Parameter s represents measure characterizing the degree of nonlinearity in the system. Therefore, the solution (180) contains (as subsequences for $s=0$) one-dimensional harmonic waves, while for $s \to 0$ one-dimensional wave packet. These two subsequences define the non-quasi-autonomous regime of the drug release process [48, 49, 55], i.e. the system should receive external energy in order to develop. For $s=1$, the solution (180) becomes one-dimensional soliton, while for $s \to 1$, one-dimensional soliton packet will be generated. The last two imply a quasi-autonomous regime (self evolving and independent [48]) for drug particle release process [48, 49, 55].

The three dimensional plot of solution (180) shows one-dimensional cnoidal oscillation modes of the concentration field, generated by similar trajectories of the drug particles (see Fig. 14). We mention that cnoidal oscillations are nonlinear ones, being described by the elliptic function cn, hence the name (cnoidal).

It is known that in nonlinear dynamics, cnoidal oscillation modes are associated with nonlinear lattice of oscillators (the Toda lattice [56]). Consequently, large time scale drug particle ensembles can be compared to a lattice of nonlinear oscillators which facilitates drug release process.

5.2.1. The correspondence between theoretical model and experimental results

In what follows we identify the field Φ from relation (180) with normalized concentration field of the released drug from micro particles.

For best correlation between experimental data and the theoretical model (for each sample) we used a planar intersection of the graph in Fig. 14 [57], in order to obtain two-dimensional plots.

The highest value of the correlation coefficient (for two data sets: one obtained from the planar intersection, the other from experimental data) will represent the best approximation of experimental data with the theoretical model.

Our goal was to find the right correlation coefficient which should be higher than 0.6–0.7, in order to demonstrate the relevance of the model we had in view. Figs. 15 show experimental and theoretical curves that were obtained through this method, where R^2 represents the correlation coefficient and η a normalized variable which is simultaneously dependent on normalized time and on nonlinear degree of the system (s parameter). Geometrically, η represents the congruent angle formed by the time axis and the vertical intersection plane.

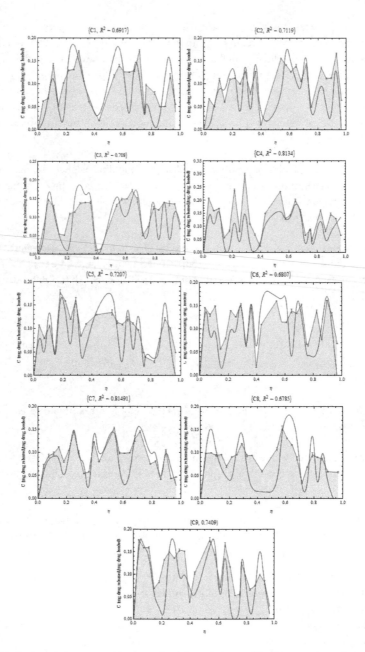

Figure 15. The best correlations among experimental and theoretical curves (blue line – experimental curve, red line – theoretical curve).

6. Conclusions

i. Scale relativistic framework is implemented by passing to a fluid-like description (the fractality of space), considering the velocity field a fractal function explicitly depending on a scale variable (the fractal geometry of each geodesic) and defining two fractal velocity fields which are fractal functions of the scale variable dt (the non-differentiability of space).

An application of these principles to the motion equation of free particles leads to the occurence of a supplementary TISE (time independent, Schrödinger-type equation) and the following interesting results :

- $\zeta(x)$ behaves like a wave function on small distances (the same magnitude as the Compton length);

- for $\gamma(x)$ a velocity potential well, $U(x)$ is quantified;

- for the harmonic oscillator case, the limit velocity χ has discrete values, and only the first value is less than the velocity of light, c;

- in the double-well velocity potential, the complex velocity $U(x)$ is again quantized, this time the levels are equally spaced at a value of $\hbar \ln 2$;

- if one takes $\phi_R = \phi_L = \pi/2$, singularities are obtained for x $x_0 - \Lambda/2$ and for $x - x_0 = \Lambda/4$ one gets minima for $U(x) = c$ in a double-well velocity potential;

- since we considered here the one-dimensional case we get the solution of a single row of rectilinear vortices, which has already been referred to as characterizing a surface of discontinuity;

- a typical bound state in a double-well has two classically allowed regions, where the velocity potential is less than the limit velocity; these regions are separated by a classically forbidden region, or barrier, where the velocity potential is larger than the limit velocity;

- for tunneling case, there is a nonzero transmission, reflection coefficient, which leads to the proof of the transport of the V field by the motion of the Newtonian fluid with velocity $U(x)$, on small distances (of the order of magnitude of Compton length).

ii. We analyzed vacuum from the Casimir cavity, considered a non-differentiable, Newtonian, 2D non-coherent quantum fluid, by writing the Navier-Stokes equations in scale relativity theory's framework. As a result the following results may be extracted:

- the (vector) velocity field V and/or the (scalar) density field ρ behave like a wave function on small distances (the same magnitude as the Compton length);

- the (vector) velocity field V and/or the (scalar) density field ρ are transported by the motion of the Newtonian fluid with velocity U, on small distances (the same magnitude as the Compton length);

Also, the entities assimilated to vortex-type objects from the Casimir cavity, initially non-coherent, become coherent due to constraints induced by the presence of walls and generate pressure along the Ox and Oy axis, thus one can stress out :

- the pressure p_y on the plates, is negative and an attractive force results, as is the case of the Casimir force;

- besides the pressure p_y acting on the plates, there must be yet another pressure, p_x, acting along the Ox axis;

- the order of magnitude of this force, $p_y \cong 6.18\,10^{10}$ N m^{-2} is the same with the value of the classical Casimir force calculation, $F_C \cong 2.08\,10^{10}$ N m^{-2};

- in the case of the Casimir cavity from inside a rectangular enclosure of sides d_1, d, the plates induce constraints along both Ox and Oy axis, and one can notice that if the two parameters m and n have close values, the force acting on the Casimir rectangle is always negative and for parameters m and n very asymmetric the force has negative and positive domains, in agreement with the calculus of regularization using the Abel-Plana formula.

iii. Using fractional calculus, the fractal "diffusion" equation give rise to Weibull relation, a statistical distribution function of wide applicability, inclusively in drug release studies. In this approach, we consider all the simultaneous phenomena involved, equivalent with complexity and fractality, offering, in this way, a physical base to this equation and for its parameters. They are functions of fractal dimension of the curves on which drug release mechanism takes place, dimension that is a measure of the complexity and nonlinear dynamics of the system, dependent on the diffusion order.

This theory offers new alternatives for the theoretical study of drug release process (on large time scale) in the presence of all phenomena and considering a highly complex and implicitly, non linear system. Consequently, the concentration field has cnoidal oscillation modes, generated by similar trajectories of drug particles. This means that the drug particle ensemble (at time large scale) works in a network of non linear oscillators, with oscillations around release boundary. Moreover, the normalized concentration field simultaneously depends on normalized time non linear system (through s parameter).

Author details

M. Agop[1], C.Gh. Buzea[2], S. Bacaita[1], A. Stroe[3] and M. Popa[4]

1 Department of Physics, Faculty of Machine Manufacturing and Industrial Management, "Gheorghe Asachi" Technical University of Iasi, Iasi, Romania

2 National Institute of Research and Development for Technical Physics, Romania

3 National College "Nicolae Balcescu", Al. I. Cuza Bvd., Braila, Romania

4 Department of Natural and Synthetic Polymers, Faculty of Chemical Engineering and Environmental Protection, "Gheorghe Asachi" Technical University of Iasi, Iasi, Roma

References

[1] L. Nottale, Fractal Space-Time and Microphysics: Towards a Theory of Scale Relativity, World Scientific Publishing, Singapore, 1993.

[2] L. Nottale, Scale Relativity and Fractal Space-Time – A New Approach to Unifying Relativity and Quantum Mechanics, Imperial College Press, London, 2011.

[3] L. Nottale, Fractals and the quantum theory of space time, Int. J. Mod. Phys. A, vol. 4, no. 19, pp. 5047-5117, 1989.

[4] L. D. Landau, E. M. Lifshitz, Fluid Mechanics, 2nd Edition, Butterworth Heinemann Publishing, Oxford, 1987.

[5] B.B. Mandelbrot, The Fractal Geometry of Nature, Freeman, San Francisco, USA, 1983.

[6] J. F. Gouyet, Physique et Structures Fractals, Masson, Paris, 1992.

[7] M.S. El Naschie, O. E. Rössler, I. Prigogine, Quantum Mechanics, Diffusion and Chaotic Fractals, Elsevier, Oxford, 1995.

[8] P. Weibel, G. Ord, O. E. Rösler, Space Time Physics and Fractality, Springer Dordrecht, 2005.

[9] M. Agop, N. Forna, I. Casian Botez, C. Bejenariu, New theoretical approach of the physical processes in nanostructures, J. Comput. Theor. Nanosci., vol. 5, no. 4, pp. 483-489, 2008.

[10] I. Casian-Botez, M. Agop, P. Nica, V. Paun, G.V. Munceleanu, Conductive and convective types behaviors at nano-time scales, J. Comput. Theor. Nanosci., vol. 7, no. 11, pp. 2271-2280, 2010.

[11] L. Nottale, Scale-relativity and quantization of the universe. I. Theoretical framework, Astron. Astrophys., vol. 327, no. 3, pp. 867-889, 1997.

[12] P.D.McCormack, L. Crane, Physical Fluid Mechanics, Academic Press, London-NewYork, 1973.

[13] M. Agop, P.D.Ioannou, C.Gh.Buzea, P.Nica, Hydrodynamic formulation of scale relativity theory and unified superconductivity by means of a fractal string, Physica C, vol. 390, no. 1, pp. 37-55, 2003.

[14] M. Agop, P.E. Nica, P.D. Ioannou, A. Antici, V.P. Paun, Fractal model of the atom and some properties of the matter through an extended model of scale relativity, The European Phys. J. D, vol. 49, no. 2, pp. 239-248, 2008.

[15] V. P. Maslov, M. V.Fedoriuk, Semiclassical Approximation in Quantum Mechanics, Reidel, Dordrecht, 1981.

[16] W. Nolting, Quantenmechanik-Methoden und Anwendungen., Grundkurs Theoretische Physik Springer, 2004.

[17] C. S. Park, M. G. Jeong, S.-K. Yoo, D.K. Park, arXiv:hep-th/9808137v1, 1998.

[18] R. Landauer, Irreversibility and Heat Generation in the Computing Process, IBM J. Res. Dev., vol. 5, no. 3, pp. 183-192, 1961.

[19] S. Lloyd, Use of mutual information to decrease entropy: Implications for the second law of thermodynamics, Phys. Rev. A, vol. 39, no. 10, pp. 5378-5386, 1989.

[20] C.H. Bennett, The thermodynamics of computation—a review, Int. J. Theor. Phys., vol. 21, no. 12, pp. 905-940, 1982.

[21] E. Fredkin, T. Toffoli, Conservative logic, Int. J. Theor. Phys., vol. 21, no. 3-4, pp. 219-253, 1982.

[22] W.H. Zurek, Algorithmic randomness and physical entropy, Phys. Rev. A, vol. 40, no. 8, pp. 4731-4751, 1989.

[23] C.R. Calidonna, A. Naddeo, Towards reversibility in a JJL qubit qualitative model by means of CAN2 paradigm, Phys. Lett. A, vol. 358, no. 5-6, pp. 463-469, 2006.

[24] R. Alicki, M. Horodecki, P. Horodecki, R. Horodecki, Thermodynamics of Quantum Information Systems — Hamiltonian Description, Open Sys. & Information Dyn., vol. 11, no. 3, pp. 205-217, 2004.

[25] V. S. Popov, B. M. Karnakov, V. D. Mur, On matching conditions in the WKB method, Phys. Lett. A, vol. 210, no. 6, pp. 402-408, 1996.

[26] C. Eltschka, H. Friedrich, M. J. Moritz, J. Trost, Tunneling near the base of a barrier, Phys. Rev. A, vol. 58, no. 2, pp. 856-861, 1998.

[27] M. J. Moritz, Tunneling and reflection of long waves, Phys. Rev. A, vol. 60, no.2, pp. 832-841, 1999.

[28] S.K. Lamoreaux, Demonstration of the Casimir force in the 0.6 to 6μm range, Phys. Rev. Lett., vol. 78, no. 1, pp. 5-8, 1997.

[29] M. Bordag, U. Mohideen, V.M. Mostepanenko, New developments in the Casimir effect, Phys. Rep., vol. 353, no. 1-3, pp. 1-206, 2001.

[30] K. Milton, The Casimir effect: recent controversies and progress, J. Phys. A, vol. 37, no. 38, pp. R209-R277, 2004.

[31] V.V. Nesterenko, G. Lambiase, G. Scarpetta, Calculation of the Casimir energy at zero and finite temperature: Some recent results, Riv. Nuovo Cimento, vol. 027, no. 06, pp. 1-74, 2004.

[32] S.K. Lamoreaux, The Casimir force: background, experiments and applications, Rep. Prog. Phys., vol. 68, no. 1, pp. 201-236, 2005.

[33] G. Barton, Perturbative Casimir energies of dispersive spheres, cubes and cylinders, J. Phys. A: Math. Gen., vol. 34, no., pp. 4083-4114, 2001.

[34] K. A. Milton, The Casimir Effect: Physical Manifestations of Zero-Point Energy (chapter 2), World Scientific, 2001.

[35] L. E. Ballentine, Quantum Mechanics (chapter 19), Prentice-Hall, 1990.

[36] C. Itzykson, J. B. Zuber, Quantum Field Theory (chapter 3), McGraw-Hill, 1985.

[37] K. Huang, Quantum Field Theory (chapter 5), John Wiley, 1998.

[38] M. Ignat, N. Rezlescu, C.Gh.Buzea, C. Buzea, About the pair breaking-time in super-conductors, Phys. Lett. A vol. 195, no. 2, pp. 181-183, 1994.

[39] C. Gh. Buzea, M. Agop, N. Rezlescu, C. Buzea, T. Horgos, V. Bahrin, The Time of Diffusion and Infinite Conductivity of High-Tc Superconductors, Phys. Stat. Sol.(b), vol. 205, no. 2, pp. 595-602, 1998.

[40] M. Agop, C. Gh. Buzea, N. Rezlescu, C. Buzea, C. Marin, Wave guide perturbative solutions for the Ginzburg–Landau equation.: Infinite conductivity and discrete values of the critical temperature in superconductors, Physica C, vol. 313, no. 3-4, pp. 219-224, 1999.

[41] F. Bowman, Introduction to elliptic functions with applications, English University Press London, 1961.

[42] O. Mayer, Special issues in the theory of the functions with one complex variable, vol. II, Academic Press Bucharest, 1990.

[43] M. Agop, V. Griga, C. Buzea, C. Stan, D. Tatomir, The uncertainty relation for an assembly of Planck-type oscillators. A possible GR-quantum mechanics connection, Chaos, Solitons & Fractals, vol. 8, no. 5, pp. 809-821, 1997.

[44] S. Titeica, Quantum Mechanics, Academic Press Bucharest, 1984.

[45] M. Agop, N. Rezlescu, G. Kalogirou, Nonlinear Phenomena in Materials Science, Graphics Art Publishing House Athens, 1999.

[46] G. Burns, High-Temperature Superconductivity, Academic Press San Diego, 1992.

[47] M. Bordag, G. L. Klimchitskaya, U. Mohideen, V. M. Mostepanenko, Advances in the Casimir Effect., Oxford Univ. Press, 2009.

[48] M. Agop, N. Forna, I. Casian Botez, C. Bejenariu, New theoretical approach of the physical processes in nanostructures, J. Comput. Theor. Nanosci., vol. 5, no. 4, pp. 483-489, 2008.

[49] I. Casian-Botez, M. Agop, P. Nica, V. Paun, G.V. Munceleanu, Conductive and convective types behaviors at nano-time scales, J. Comput. Theor. Nanosci., vol. 7, no. 11, pp. 2271-2280, 2010.

[50] L. Jude, Mathematics physics equations. Theory and applications, Matrix Rom Publishing Bucharest, 2010

[51] K. B. Oldham, J. Spanier, The Fractional Calculus: Theory and Applications of Differential and Integration to Arbitrary Order, Dover Publications New York, 2006.

[52] A. A. Kilbas, H. M. Srivastava, J. J. Trujilto, Theory and Applications of Fractional Differential Equations, Elsevier Armsterdam, 2006.

[53] S. Bacaita, C. Uritu, M.Popa, A. Uliniuc, C. Peptu, M. Agop, Drug release kinetics from polymer matrix through the fractal approximation of motion, Smart Materials Research, article ID 264609, doi:10.1155/2012/264609, 2012.

[54] A. J. Lichtenberg, Phase-Space Dynamics of Particle, John Wiley and Sons Inc. New York, 1969.

[55] S. Popescu, Actual issues in the physics of self-structured systems, Tehnopress Publishing, Iasi, Romania, 2003.

[56] M. Toda, Theory of Nonlinear Lattices, Springer, Berlin, 1989.

[57] S. Bacaita et. al., Nonlinearities in Drug Release Process from Polymeric Microparticles: Long-Time-Scale Behaviour, Journal of Applied Mathematics, vol. 2012, article ID 653720, 2012, doi:10.1155/2012/653720.

Permissions

The contributors of this book come from diverse backgrounds, making this book a truly international effort. This book will bring forth new frontiers with its revolutionizing research information and detailed analysis of the nascent developments around the world.

We would like to thank Professor Paul Bracken, for lending his expertise to make the book truly unique. He has played a crucial role in the development of this book. Without his invaluable contribution this book wouldn't have been possible. He has made vital efforts to compile up to date information on the varied aspects of this subject to make this book a valuable addition to the collection of many professionals and students.

This book was conceptualized with the vision of imparting up-to-date information and advanced data in this field. To ensure the same, a matchless editorial board was set up. Every individual on the board went through rigorous rounds of assessment to prove their worth. After which they invested a large part of their time researching and compiling the most relevant data for our readers. Conferences and sessions were held from time to time between the editorial board and the contributing authors to present the data in the most comprehensible form. The editorial team has worked tirelessly to provide valuable and valid information to help people across the globe.

Every chapter published in this book has been scrutinized by our experts. Their significance has been extensively debated. The topics covered herein carry significant findings which will fuel the growth of the discipline. They may even be implemented as practical applications or may be referred to as a beginning point for another development. Chapters in this book were first published by InTech; hereby published with permission under the Creative Commons Attribution License or equivalent.

The editorial board has been involved in producing this book since its inception. They have spent rigorous hours researching and exploring the diverse topics which have resulted in the successful publishing of this book. They have passed on their knowledge of decades through this book. To expedite this challenging task, the publisher supported the team at every step. A small team of assistant editors was also appointed to further simplify the editing procedure and attain best results for the readers.

Our editorial team has been hand-picked from every corner of the world. Their multi-ethnicity adds dynamic inputs to the discussions which result in innovative outcomes. These outcomes are then further discussed with the researchers and contributors who give their valuable feedback and opinion regarding the same. The feedback is then collaborated with the researches and they are edited in a comprehensive manner to aid the understanding of the subject.

Apart from the editorial board, the designing team has also invested a significant amount of their time in understanding the subject and creating the most relevant covers. They scrutinized every image to scout for the most suitable representation of the subject and create an appropriate cover for the book.

The publishing team has been involved in this book since its early stages. They were actively engaged in every process, be it collecting the data, connecting with the contributors or procuring relevant information. The team has been an ardent support to the editorial, designing and production team. Their endless efforts to recruit the best for this project, has resulted in the accomplishment of this book. They are a veteran in the field of academics and their pool of knowledge is as vast as their experience in printing. Their expertise and guidance has proved useful at every step. Their uncompromising quality standards have made this book an exceptional effort. Their encouragement from time to time has been an inspiration for everyone.

The publisher and the editorial board hope that this book will prove to be a valuable piece of knowledge for researchers, students, practitioners and scholars across the globe.

List of Contributors

C. M. Arizmendi and O. G. Zabaleta
Facultad de Ingeniería, Universidad Nacional de Mar del Plata, Argentina

Peter Enders
University of Applied Sciences, Wildau, Königs Wusterhausen, Germany

Flavia Pennini
Departamento de Física, Universidad Católica del Norte, Antofagasta, Chile
Instituto de Física La Plata–CCT-CONICET, Fac. de Ciencias Exactas, Universidad Nacional de La Plata, La Plata, Argentina

Sergio Curilef
Departamento de Física, Universidad Católica del Norte, Antofagasta, Chile

GianCarlo Ghirardi
Professor Emeritus, University of Triest, Italy

M. D. Bal'makov
Faculty of Chemistry, St. Petersburg State University, Staryi Peterhof, Universitetskii Pr. 26, St. Petersburg, Russia

Rodolfo O. Esquivel
Departamento de Química, Universidad Autónoma Metropolitana-Iztapalapa, México, D.F., México
Instituto "Carlos I" de Física Teórica y Computacional, Universidad de Granada, Granada, Spain

Catalina Soriano
Laboratorio de Química Computacional, FES-Zaragoza, Universidad Nacional Autónoma de México, C.P. 09230 Iztapalapa, México, D.F., México

Carolina Barrientos
Facultad de Bioanálisis-Veracruz, Universidad Veracruzana, Lab. de Química y Biología Experimental, Veracruz, México

Jesús S. Dehesa
Departamento de Física Atómica, Molecular y Nuclear, Universidad de Granada, Granada, Spain
Instituto "Carlos I" de Física Teórica y Computacional, Universidad de Granada, Granada, Spain

José A. Dobado
Grupo de Modelización y Diseño Molecular, Departamento de Química Orgánica, Universidad de Granada, Granada, Spain

Moyocoyani Molina-Espíritu and Frank Salas
Departamento de Química, Universidad Autónoma Metropolitana-Iztapalapa, México, D.F., México

Bjørn Jensen
Vestfold University College, Norway

Jonathan Bentwich
Brain Perfection LTD, Israel

Nelson Flores-Gallegos
Universidad Autónoma Metropolitana-Iztapalapa, México

M. Agop and S. Bacaita
Department of Physics, Faculty of Machine Manufacturing and Industrial Management, "Gheorghe Asachi" Technical University of Iasi, Iasi, Romania

C.Gh. Buzea
National Institute of Research and Development for Technical Physics, Romania

A. Stroe
National College "Nicolae Balcescu", Al. I. Cuza Bvd., Braila, Romania

M. Popa
Department of Natural and Synthetic Polymers, Faculty of Chemical Engineering and Environmental Protection, "Gheorghe Asachi" Technical University of Iasi, Iasi, Romania

Printed in the USA
CPSIA information can be obtained
at www.ICGtesting.com
JSHW011445221024
72173JS00004B/947

9 781632 381620